AF601556

Aquarium Feed

Grasshopper Minilivestock

The Author

Dr. Arijit Ganguly was born and brought up in the serene environment of Santiniketan, West Bengal, India; the abode of Gurudev Rabindranath Tagore, the first Nobel Laureate from Asia. Gradually, he completed his education in Zoology and obtained his PhD from Visva-Bharati University under the efficient guidance of Prof. Parimalendu Haldar and Prof. Dipak Kr. Mandal. Later, he joined Rajiv Gandhi University, Arunachal Pradesh as a postdoctoral Research Associate where he was trained by Prof. Debangshu Narayan Das.

Presently, Dr. Ganguly is working as an Assistant Professor of Zoology in the Achhruram Memorial College of the beautiful district Purulia in West Bengal, India. Since last eight years Dr. Ganguly is working in the field of utilisation of insects as food for livestock. He has published a good number of research articles in the journals of international repute.

Aquarium Feed

Grasshopper Minilivestock

Dr. Arijit Ganguly

2017

Scholars World

A Division of

Astral International Pvt. Ltd.

New Delhi – 110 002

Cataloging in Publication Data--DK
Courtesy: D.K. Agencies (P) Ltd. <docinfo@dkagencies.com>

Ganguli, Arijit, author.
Aquarium feed : grasshopper minilivestock / Dr. Arijit Ganguly.
pages cm
Includes bibliographical references.
ISBN 978-93-86071-42-2 (International Edition)

1. Grasshoppers--Feeding and feeds--India. 2. Aquarium fishes--Feeding and feeds--India. 3. Elemental diet. I. Title.

SB945.G65G36 2017 DDC 638.57260954 23

Published by : **Scholars World**
A Division of
Astral International Pvt. Ltd.
– ISO 9001:2015 Certified Company –
4736/23, Ansari Road, Darya Ganj
New Delhi-110 002
Ph. 011-4354 9197, 2327 8134
E-mail: info@astralint.com
Website: www.astralint.com

Preface

Ornamental fish industry is a profitable industry in various part of the world. Even a decade back keeping these fishes were considered as a mere hobby but from the last decade various researchers reported about their positive effect on mental health. This has established ornamental fish industry as a need for modern day civilisation as they have been mainly reported to be able to control stress. In this scenario the need of these fishes are growing day by day. But the ornamental fish farmers are facing difficulty in raising these fishes because of a very high cost of their feed. Consequently the production cost is very high rising and so is the cost of the fishes in the market. Moreover it has been established that like the other fishes, the best conventional protein source for the ornamental fish industry also is "fish meal" itself. As fish is an important food for many other livestock and human as well, there is an overexploitation of fish meal along with the ever increasing human population. Thus a high "demand : supply" ratio of the fish meal is an additional problem for this industry. In this context insects have the potential to play an important role as an alternative protein supplement in the ornamental fish diets.

Although people of different countries, especially various ethnic groups consume insects from time immemorial, this highly potential protein source has hardly got any attention of researchers. Apart from some scattered information regarding evaluation of insects as food for human and livestock the scientific community has remained almost silent for a long time.

However, after the Food and Agriculture Organisation (FAO) of the United Nations (UN) published their book "Edible insects: Future prospects for food and feed security" in 2013, research on food insects has got some real attention worldwide.

Research on food insect is in its infancy in India. It is quite natural that there is no complete work on nutritional evaluation of insects, estimation of their farming potential, feed formulation and feeding trials with insect supplemented diets for livestock. Hence the present book explores how grasshoppers could play a significant role in the formulated diets for a model ornamental fish "black molly". The main goal of this book is to make people aware of food insects and their nutritional potential as a livestock feed. Researchers may find this book interesting if they want to work in the field of fish nutrition, food insects, ornamental fisheries etc.This book might be also useful for fish farmers, ornamental fish traders and all those people having interest in aquarium fishes.

Dr. Arijit Ganguly

Contents

List of Figures

Chapter 1

Prelude: The Journey Begins

1.1. Introduction

Keeping ornamental fishes as pet is an ancient practice that dates back early 17th century in Europe, while in Asia it is more than a thousand year. However, from the very beginning ornamental fishes were considered only for their aesthetic value. Situation was very similar even a decade back or two. Keeping them at home in an aquarium was a mere hobby and was considered as an extravagance of some rich people. However, things started to change from the beginning of 1990s. Modern research has revealed that these beautiful fishes also have tremendous therapeutic value especially in mental, nervous and stress related problems of people. As for example ornamental fishes can reduce anxiety, maintain blood pressure, and improve sleep. These fishes also have a positive effect on Alzheimer's patients. According to some researchers, appetite of the Alzheimer's patients could have been improved and the patients showed less physically aggressive and disruptive behaviours after an aquarium was placed in their room. The stimulating combination of colour, sound, and the varying movement of the fish not only held their attention for a relatively long time but they were also more relaxed and alert; for some, even short-term memory was observed to be stimulated. Recently another study reported that the ornamental fishes also have tremendous positive effect on the children diagnosed with attention deficit hyperactivity disorder (ADHD). In this scenario, the need for the ornamental fish industry is growing day by day.

Singapore and USA are the two major countries providing more than 50 per cent of the total amount supplied throughout the globe annually. Among the common ornamental fishes, rearing the tropical live bearers such as guppies, swordtails and mollies is an established industry in countries such as Singapore, which is the largest ornamental fish supplier in the world. In Asia, countries like Sri Lanka, China and India are emerging as suppliers of ornamental fishes very recently. But like every aquaculture industries worldwide, the ornamental fisheries industry is also facing the same basic problems, and the problem is almost the same globally.

It is apparent that nutrition is a matter of great importance in aquaculture industry worldwide; a better product quality and optimum growth of this industry can be achieved by using appropriate feeds, and aqua-feed is the single largest input in aquaculture. At present, India is using 5.4 million tons of aqua-feed annually and the demand may rise over 10 million tons by 2020 in order to sustain the annual production of fin-fish. One important ingredient used in the formulation of commercial aquaculture feed is fishmeal, which has high protein quality and palatability. But the problem is the limitation in the resource and production of fish meal in our country.

India has 2.02 million sq kilometers of exclusive economic zone (EEZ). In 1992-93 the annual fish production of total 4.04 million tons was estimated where 2.24 million tons came from the marine sector and1.8 million tons came from the inland sector. It is estimated that India has the potential of fish production capacity of 7.9 million tons (3.9 million tons in marine sector and 4.0 million tons in the inland sector). A little part of the total fish landing is used to produce good quality fish meal and the lion's share of the produced fishmeal goes for the production of poultry and livestock feed. Thus the demand of fish meal is growing day by day with the growing livestock industry to compete with the ever increasing human population. That is why fish meal is being overexploited worldwide though the resource is limited. In this context there is an urgent need to find out alternative protein supplement that is almost equivalent to fish meal especially in terms of protein and energy.

Ornamental fishes have traditionally been fed live feed, many of which are arthropods. As for example, a daily supplementation of *Daphnia* sp. as live feed to sword tail (*Xiphophorus helleri*) broodstock maintained on an artificial flake diet results in a significant increase in fecundity showing more rapid growth, a higher number of embryos and an improved feed conversion ratio. In fresh water ornamental fish culture, Moina used to be the most common live feed organism for feeding young fish in the industry. Moreover some insects that explore into water or near water are an easy meal to many carnivorous fishes in nature. The archer fish *Toxotes jaculatrix* is a highly specialized predator, which feeds on insects that it shoots down from above the surface with a jet of water. In this scenario insects could be an attractive alternative protein source to feed the ornamental fishes.

1.2. Why Insects?

1.2.1. Entomophagy in different Cultures

The term Entomophagy has been originated from Greek words *éntomos*, which means "insect" and *phãgein*, which means "to eat". According to the definition entomophagy is the consumption of insects as food. Insects have played an important part in the history of human nutrition in Africa, Asia and Latin America. Worldwide, nearly 1,700 insect species have been reported to be used as human food in 124 countries, including 549 species only in Mexico. Among the edible insects four insect orders predominate in rank sequence: Coleoptera, Hymenoptera, Orthoptera and Lepidoptera, accounting for 80 percent of the species eaten. Geographically, the Americas and Africa are identified as the prime insect consuming zones, where respectively 679 and 524 edible insect species have been recorded. On the other hand when the Pacific countries were combined with Asian countries, the region registered more than 500 insect species consumed as food in total 43 countries. The total number of species eaten in Asia might be considerably higher than this number, as research on the subject appears to have been less rigorous in Asia and the Pacific compared to the works conducted and published in Africa and the Americas. Among different ethnic groups, Yukpa people of Colombia and Venezuela prefer certain of their traditional insect foods to fresh meat, as do the Pedi of South Africa. In Australia amongst the native aborigines – Witchery grub, a caterpillar of wood moth is considered as a delicacy. The larva of *Cirina forda* (Lepioptera: Saturniidae) is served as snacks or taken with carbohydrate food in Southern Nigeria. In Mexico, the eggs of certain large aquatic bugs are regularly sold in the city markets. The Mexicans sink sheets of matting under water upon which millions of eggs are laid by the bugs and then dried, placed in sacks, sold by the kilogram and used for making cakes and other foods. In Jamaica, the local people consider a plate of crickets a compliment to the most distinguished guests. In Australia, the natives collect quantities of the bugony moth (*Agrotis infusa*) in bags, roast and eat them and claim that they taste like nuts. The manna (surgery honey dew) excreted by aphids and scale insects is used as a sweet by peasants in Turkey, Iraq and Iran. Locusts are eaten with gusto (fried and seasoned with salt and pepper or in cakes) in Arabia, Persia and Madagascar. In Africa among various tribes, boiled termites are the most popular insect food. In some regions of South America and Africa, termite colonies are often staked out as the private property of individuals or groups. The larvae of the palm weevil are extracted from the palm trees and eaten.

From different states of India also edible insects under different orders and families have been reported by various workers. Among them Orthoptera, Hemiptera, Coleoptera, Odonata are the most common. Therefore, we can say that edible insects are eaten all over the world. Around 3,041 ethnic groups have been registered till date, to whom this naturally renewed resource is a part of daily diet. The insect species that are eaten vary in relation to the region and to the season, as well as according to their abundance. Selection is made over time, based on size, flavour, storage capacity, energy used in their capture, time to obtain them, demand, colour, the best stage, time of acquisition, and the nutritional value, which is evaluated on the basis of the degree of satisfaction and wellbeing.

1.2.2. Nutritional Value of Edible Insects

Insects are rich in protein. Protein contents of edible insects are mainly reported from African and South American countries since a long time by various workers. In Asia from China only100 kinds of edible insects have been analysed till date. At the egg, larva, pupa and adult stages, the raw protein content is generally 20–70 per cent. The raw protein content of the Ephemeropteran larva has been found to be 66.26 per cent, followed by Odonata larva, 40–65 per cent; Homoptera larva and eggs, 40–57 per cent; Hemiptera, 42–73 per cent; Coleoptera larva, 23–66 per cent; and Lepidoptera, 20–70 per cent. The protein content of Formicidae in Hymenoptera is 38–76 per cent, while the same of Apidae and Vespidae of Hymenoptera is 15–70 per cent. According to analysed data, the protein content of insects is obviously higher than in most plants; sometimes it is also higher than that of chicken eggs, meat and fowl.

According to a report from South-Western Nigeria, twelve species have been found to contain more than 20.0 per cent crude protein, where the highest amount of protein (29.62 per cent) was observed in *Analeptes trifasciata*. From Mexico, the highest amount of protein was found in the red-legged locusts (75.30 per cent),whereas agave worms had a little lower amount (30-37 per cent), however, most of them contained around 40-60 per cent crude protein. Protein is composed of 20 amino acids which are necessary for nutrition. The essential amino acids for human adult nutrition are the following: isoleusine, lysine, methionine, phenylalanine, threonine, tryptophan, valine and histidine. The amino acid contents of many more edible insects have been reported from different part of the world, such as: caterpillars of Zaire, edible insects from Mexico, mormon crickets of Colorado, USA and edible insects from China. The Chinese edible insects have shown that they contain all the necessary amino acids. Studies on mormon cricket and house cricket showed that they were deficient in methionine. In addition, some lepidopteran species were also found to be deficient in methionine, cysteine and possibly lysine. Therefore it could be concluded that insects might be unsatisfactory as the sole source of dietary protein due to limiting amino acids, but would be extremely beneficial as a supplement.

Fat is one of the main components of any living organism. Various authors have shown that many edible insects are rich in fat, and their fat content is higher at the larval and pupa stages, while at the adult stage, the fat content is relatively lower. For example, the fat content of *Oxya chinensis* adults (Orthoptera) is only 2.2 per cent but yet, the fat content of edible insects is mostly in between 10–50 per cent. Some larva and pupa of Lepidoptera have higher fat contents, such as *Pectinophora gossypeilla* (49.48 per cent) and *Ostrinia furnacalis* (46. 08 per cent). The fatty acid content of edible insects is different from animal fat, as insects have a higher essential fatty acid content, which the human body needs. These fatty acids are found in the larva and pupa of *Dendrolimus houi*, larva of *Musca domestica*, and in some ants. Therefore, the fat of edible insects has good nutritive value. Fat can store and supply energy. For example, the calorific value of 761 kcal/100g is reported in the winged sexual forms of the African termite,

Macrotermes falciger, while the winged forms of another African species, *Macrotermes subhyalinus* were found to contain 613kcal /100g (dry weight), whereas a calorific value of 611kcal/100g is reported for the caterpillar *Anaphe venata* in Nigeria. Additionally, twenty three species of caterpillars in Zaire, mostly of Saturniidae family, were found to average 457 kcal/100 g dry weights, ranging from 397 to 543 kcal/100g. Recent analyses of 94 edible insect species of Mexico also yielded high fat and calorific values. Regarding carbohydrate content edible insects are lagging behind. Insects could be a good source of minerals and vitamins also. Among the minerals sodium, potassium, calcium, zinc, iron and magnesium are prominent in literature. Orthoptera, Lepidoptera and Hymenoptera are the orders with lowest variation in mineral content. The edible insects are mostly low in sodium, and sometimes high in calcium, and rich in zinc, iron, potassium and magnesium: as for example, termites are rich in calcium and sulphur, whereas grasshoppers are rich in iron and zinc. Among vitamins, insects have been found to be rich in B group, such as niacin, riboflavin and thiamine.

From the literature of nearly half a century on the chemical analysis of edible insects, we could now be sure that these insects are rich in protein (with the presence of essential amino acids), as well as vitamins, minerals and fats. Hence, these are not only suitable for those who already consume them from the time immemorial but it might be an answer to the future food scarcity as the world gradually running out of its conventional protein source. However, for a constant supply for consumption, insect farms should be built to rear them in mass scale. The idea of insect farm is promising because many insects naturally produce a huge annual biomass— for example, Argelia Locusts and *Schistocerca* produce 9 tons of biomass per year and in Mexico *Sphenarium* sp. produces more than 10 tons of biomass per year. In this scenario the concept of insect minilivestock (intentional cultivation of arthropods for human food) is now emerging in animal husbandry. The major advantage being that they do not have to be fed on grains thus saving many crop species for human consumption. It is also considered to be much more eco-friendly than traditional livestock.

1.2.3. Insect Farms: More Eco-friendly than the Traditional Livestock

The livestock sector contributes up to 18 per cent of total anthropogenic greenhouse gas (GHG) emissions. It has also been estimated that the global contribution to GHG emissions by the animal sector are: 9 per cent for CO_2 (fertilizer production for feed crops, on farm energy expenditures, feed transport, animal product processing, animal transport, and land use changes), 35–40 per cent for CH_4 (enteric fermentation in ruminants and from farm animal manure) and 65 per cent for N_2 (farm manure and urine) based on their studies on Life Cycle Analysis (LCA) that takes the entire production process of animal products into account. Livestock is also associated with environmental pollution due to ammonia (NH_3) emissions from manure and urine, leading to nitrification and acidification of soil. The main source of gaseous NH3 is bacterial fermentation of uric acid in poultry manure and bacterial fermentation of urea in mammals. Livestock is estimated to be responsible for 64 per cent of all anthropogenic NH_3 emissions. Although not considered as a GHG, NH3 can indirectly contribute to N_2O emission. Besides these environmental problems the livestock sector faces challenges regarding resistance to antibiotics, zoonosis and animal welfare. In this scenario there is an urgent need to find alternatives for conventional sources of animal protein. To combat this situation, the concept of "minilivestock", for instance edible insects, have been suggested as an alternative source of animal protein. A handful of research works opined that production of animal protein in the form of edible insects supposedly has a lower environmental impact than the conventional livestock. Recently the greenhouse gases produced per kilogram of insect product have been quantified by a research group in Netherlands. The gases concerned were methane (CH_4) and nitrous oxide (N_2O). The results demonstrated that insects produce much smaller quantities of greenhouse gases than conventional livestock such as cattle and pigs. For example, a pig produces as much as ten to hundred times of greenhouse gases per kilogram compared with mealworms (*Tenebrio molitor*). Emissions of ammonia (which causes the acidification and eutrophication of groundwater) also appeared to be significantly lower, where a pig was estimated to produce as much as eight to twelve times ammonia per kilogram of growth compared to crickets, and up to fifty times more than locusts.

1.2.4. Rejection by the Westernized Culture: Major Hurdle for Entomophagy

Up to this level we have seen that insects are nutritionally rich and insect farms produce much less greenhouse gases than the conventional livestock. Moreover human entomophagy is accepted as a normal part of the diet in many continents from time immemorial. Despite its huge beneficial roles, eating insects is still a cultural taboo in many countries. Perhaps the general rejection of entomophagy is only an issue of marketing to counter a popular conception that insect food is for the poor. Typically the western repugnance is adding a problem, as the modern era of globalization has been seen to adopt a universal cultural system based largely on western values and habits including changes in food customs. Though there is a major attitudinal barrier to the use of insects as human food in westernised societies, interestingly most of them inadvertently consume insects because of the levels permitted in food products. For example, the allowable amounts of insects per 100 g of processed food products are 80 insect fragments in chocolate, 60 aphids, thrips or mites in frozen broccoli, 100 insect fragments in macaroni and other noodle products, 60 insect fragments in peanut butter and 150 insect fragments in wheat flour in the USA. Nevertheless it is believed that education on cultural, nutritional and ecological issues associated with entomophagy can partly overcome the aversion towards insects. According to a survey on the effect of "bug banquets" on attitudes towards insect consumption, results were polarized: people found it either disgusting or interesting. However, many were open to the idea if it was necessary for survival.

1.2.5. Insects as Livestock Feed: An Alternative to Human Consumption

If somebody does not relish the prospect of eating insects themselves, then, perhaps the concept of considering insects as analternative protein source for livestock animals is more acceptable. Various workers have incorporated insects in formulated diets for livestock, mainly for poultry and fish. Mormon cricket has been incorporated replacing soybean meal as major source of protein in practical chick diet. A variety of insects have been shown equivalent or superior to soybean meal as a high protein source for chick growth. There were no significant differences in weight gain, food consumption, food conversion, carcass quality or palatability of birds when *Tenebrio molitor* was substituted for soybean meal in the diets for young chickens. Similar finding had been reported when poultry birds were fed with *Anabrus simplex, Acheta domesticus, Bombyx mori, Apis mellifera, Alphitobius diaperinus, Tribolium castaneum* and termites. Interestingly, no significant difference in weight gain has been observed between one group of chicks fed with house fly pupae and the other fed a fully balanced diet; moreover no adverse effect on carcass quality and taste of the birds was observed. In a recent work high protein diets were formulated with the Chinese grasshopper *Acrida cineria* that proved the insect to be an acceptable feed for broiler without any adverse effect on weight gain, feed intake or "gain: feed" ratio. A review of literature indicated that insects in various developmental stages have received considerable attention as alternate sources of protein for fishes also. Aerial insects have been attracted by light to be used as supplemental food for channel catfish and bluegill sunfish. Maggot meal has been reported to possess a great potential as an alternative protein source for fish. It has been also demonstrated that blue tilapia, grow satisfactorily over a 10-week period when fed with diets containing 50, 75 and 100 per cent soldier fly larvae. The effect of inclusion of grasshopper meal on the growth, feed conversion ratio and survival of *Clarias gariepinus* fingerlings was investigated by another group of workers. The findings show that the diet having 10 per cent grasshopper meal and 30 per cent fish meal is best for growth and food utilisation, while the best survival of 100 per cent was observed in the diet containing 30 per cent grasshopper meal and 10 per cent fish meal. Looking at these very encouraging results one should admit that insects could be a potential alternative protein source in formulated supplementary diets for livestock animals. Among the edible insects acridids might have a great future to be established as a sustainable mini-livestock. Let's first have a look what acridids are.

1.3. Acridids a General Account

Acridids are short-horned grasshoppers of the Family Acrididae under the Order Orthoptera. Majority of these insects are terrestrial and usually found feeding on all types of shrubs, herbs, agricultural crops. Acridids are almost cosmopolitan in distribution except some parts of polar regions and some oceanic islands. Very few species have global distributions. Three special types of vegetation namely Savannah, Tropical rain forest and Alpine forests foster grasshopper population well. The length of these insects ranges from 15-70mm. Its colour may be yellow, yellowish green, leafy green, brown or grey. The body of grasshopper is slender, elongated and exhibit perfect bilateral symmetry. The body is covered by an exoskeleton called chitin and differentiated into head, thorax and abdomen. Thorax contains three pairs of leg; the first two pairs are ordinary walking legs while the third pair is long, stout and specialized for jumping purpose. After copulation the gravid female lay eggs in soil in egg pods. Egg incubation period varies greatly on various ecological factors. After incubation nymphs are hatched out from egg pods, fed on nearby vegetation and develop. All the eggs laid are neither hatched as nymphs nor do all the nymphs grow to reach the adult stage. Grasshoppers are hemimetabolous insects and pass through 4-8 nymphal stages before reaching fledgling stage (0 day adult). Adults get maturity, copulate and lay egg pods again. Thus its lifecycle completes. Some of the grasshoppers have great role in weed control. They have attracted the attention of economic entomologists primarily because of their main dependence on vegetations and agricultural plantation for subsistence. Being primary consumers grasshoppers play an important role in the food chain of an ecosystem. They are natural food of birds. A US biological survey investigating stomach content of 15 wild species of birds revealed that more than 50 per cent of their annual consumption comprises of insects only.

1.4. Why Acridids?

Acridids *i.e.* locusts and grasshoppers have a high nutritional value and can be used to formulate good quality feed for livestock. The protein content of grasshopper usually varies from 52.1 per cent to 77.1 per cent. Moreover they have a potential to produce a huge biomass annually and some of them are also easy to culture.

1.5. Lacunae

We could now summarise that many of the insect species not only provide high quality proteins but also contains significant amount of minerals and vitamins. Moreover if the concept of entomophagy looks bizarre to those who are not acquainted with eating these creatures, they can use these nutritionally rich sources to feed their livestock animals.

Consequently it would be ideal if some of the species are selected according to nutritional content and cultural acceptance and cultivated in mass scale in controlled conditions. Thus insect farms could yield a huge biomass to formulate feed. As the maintenance cost would be low, they could be easily supplied for the livestock consumption purposes. Moreover they would be much more eco-friendly as it has been reported that insects are a better alternative for the production of animal protein from the perspective of GHG and NH_3 emissions compared to the conventional livestock. Among various edible insects acridids could be a good choice as they could produce a huge annual biomass and they are also nutritionally rich. Despite its remarkable potential as an alternative protein supplement, results of its feeding trial on fish models is scanty; though some reports on poultry birds do exist. However, apart from all the positive sides of insects as food a doubt does exist yet— whether the insect chitin is digestible to fish.

1.6. Insect Chitin: Is it Digestible to Fishes?

Although it is quite common to wonder that whether the insect chitin could be digestible to fishes, various studies have already indicated that a high amount of fish population has the ability to do so. As for example one study revealed that the enzyme chitinase is present in the digestive system of many fish species regardless of their dietary habits. But this enzyme is not restricted to the digestive tract only, because in another work chitinase has been detected with different pH optima and tissue distribution in fishes. In relatively recent time while working with cod (*Gadus morhua*) an upsurge of the chitinase activity was detected when the fishes were fed on crustaceans. More evidences are coming forth which provides information that some fishes even have the ability to produce endogenous intestinal chitinase. Chitinolytic activity throughout the gastro-intestinal tract and in gut wall homogenates has been also observed in cod, the highest concentrations were found to be localised in stomach and pyloric tissue. Moreover 283 strains of *Aeromonas* species isolated from the intestinal tract of common carp (*Cyprinus carpio*) that showed chitinolytic ability, which proves that more than 90 per cent of the strains in fishgut are able to produce such specific enzymes. The above mentioned evidences are enough to support the concept of eating insects that could cause no harm to the fishes, especially which are carnivorous or omnivorous species.

In the forthcoming chapters of this book we will try to explain why grasshoppers are suitable insects on the basis of nutritional quality and potential for biomass production, and how various protein rich diets could be formulated replacing fish meal by grasshopper meal for an aquarium fish.

Chapter 2

Grasshopper Species of Interest

2.1. Introduction

Whenever insects are being considered as food source for fish and the other livestock, their mass production should be kept in mind for a recurring supply to the feed producing companies. Hence it is apparent that insects like multivoltine grasshoppers are more suitable as they could yield higher annual biomass than the univoltine ones because they complete a number of life cycles annually. In this book we will mainly concentrate on two multivoltine grasshopper species, *i.e.*, *Oedaleus abruptus* (Thunberg), and *Oxya fuscovittata* (Marschall). The descriptions of the grasshopper species are depicted below:

2.2. Grasshoppers of Interest

2.2.1. *Oedaleus abruptus* (Thunberg)

These are medium sized grasshoppers with brown or white markings. Their multi-segmented antennae are nearly one third times longer than head and pronotum together. Males are about 13.5mm long and have about 39 mg dry weight, whereas females are more robust than males. These are about 19.4 mm long having 52mg dry weight. Their ventral ovipositor valve is strongly sclerotized with strongly curved apices. These are trivoltine in nature, *i.e.* they can complete 3 life cycles within a year.

Male

Female

Figure 2.1: Acridids of Interest
Oedaleus abruptus **(Thunberg)**

Distribution

India (West Bengal, Andhra Pradesh, Bihar, Goa, Himachal Pradesh, Jammu and Kashmir, Kerala, Madhya Pradesh, Odisha, Rajasthan, Tamil Nadu and Uttar Pradesh), Bangladesh, Myanmar, China, Nepal, Pakistan, Sri Lanka and Thailand.

Systematic Position: (According to Orthoptera species file version 2.0/4.0)

Kingdom- Animalia

Phylum- Arthropoda

Subphylum- Hexapoda

Class- Insecta

Subclass- Pterygota

Infraclass- Neoptera

Order- Orthoptera

Suborder- Caelifera

Infraorder- Acrididea

Superfamily- Acridoidea

Family- Acrididae

Subfamily- Odipodinae

Genus- *Oedaleus*

Species- *Oedaleus abruptus*

2.2.2. *Oxya fuscovittata* (Marschall)

This species has a shiny integument which is greenish or yellowish green in appearance. They have 26-28 segmented antennae. Males are about18.3mm long having 7.5mm long antennae. Their average dry weight is 46mg. On the contrary females are much more robust than their male counterpart. These are about 23.7mm long with about 8.0mm long antennae. They have about 65mg dry weight. This is a tetravoltine species, *i.e.*, they can complete four generations per annum.

Male

Female

Figure 2.2: Acridids of Interest
Oxya fuscovittata **(Marschall).**

Distribution

India (West Bengal, Andhra Pradesh, Jammu and Kashmir, Madhya Pradesh, Odisha, Rajasthan, and Uttar Pradesh), Western part of Pakistan, Afghanistan, South-Western part of former USSR.

Systematic Position: (According to Orthoptera species file version 2.0/4.0)

Kingdom- Animalia

Phylum- Arthropoda

Subphylum- Hexapoda

Class- Insecta

Subclass- Pterygota

Infraclass- Neoptera

Order- Orthoptera

Suborder- Caelifera

Infraorder- Acrididea

Superfamily- Acridoidea

Family- Acrididae

Subfamily- Oxyinae

Genus- *Oxya*

Species- *Oxya fuscovittata*

Chapter 3

Nutrient Composition of Grasshopper Tissues

3.1. Introduction

The ornamental fish industry is growing day by day. But their growth demands fast-growing fishes and good quality of feed with high levels of energy, protein, vitamins and essential minerals to support maximum growth of these aesthetic creatures before they can be sold. At present the major conventional protein source that is used to formulate feed is fish meal which is being over-exploited and consequently has a very high price. As a result the cost of conventional feed incurs about 50–65 per cent of the total cost of livestock production. Hence, various investigations are being carried out to find alternative non-conventional supplementary diets for these industries so that the over-exploitation of fish meal could be checked. Such alternatives may include insects, because they can be used as protein sources as well as natural pigments, vitamins or minerals. Among insects, grasshoppers like *Sphenarium histrio* contains about 77 per cent protein and *Melanoplus mexicanus* about 71 per cent, whereas the Chinese grasshopper *Acrida cinerea* has been estimated to contain more than 65 per cent crude protein and established as a potential alternative food source for poultry.

In this context in India also a preliminary study on the proximate composition and mineral content of four acridid species was conducted. However, further studies are needed to look for the most suitable species.

Moreover, for a complete picture on the suitability of any material as food there is an obvious need to further explore the contents of fatty acids, amino acids, vitamins and anti-nutritional factors. In the following sections we will explore the various nutritional and anti-nutritional factors present in the grasshopper species of our interest.

3.2. Proximate Composition of Two Grasshopper Species on Dry Matter Basis

About 67-70 per cent moisture content is present in the grasshoppers. Crude protein values range between 59 per cent and 64 per cent. The percentage of crude fat is also in between 6-7 per cent. Carbohyhrate, crude fiber and nitrogen free extract (NFE) are fairly high in *O. abruptus* (28.51 per cent, 8.29 per cent and 20.22 per cent respectively), whereas the percentage of total ash is more (5.27 per cent) in *O. fuscovittata*.

3.3. Content of Minerals in mg/kg of Dried Grasshopper Body Tissues

Presence of six minerals is reported in the grasshoppers. Among them calcium (Ca) and magnesium (Mg) are present in around 5-8mg/kg, whereas manganese (Mn) is present in the least amounts (in between 0.25-0.60 mg/kg) in these insects. Calcium, manganese and iron (Fe) are present in a little higher amount in *O. fuscovittata* (about 8.0 and 3.9 mg/kg), whereas magnesium and zinc (Zn) are present in higher amount in *O. abruptus*. However, the amount of copper (Cu) does not show any significant difference between the two insects.

3.4. Content of Fatty Acids in g/100g (per cent) of Fat in Dried Grasshopper Body Tissues

Total eight fatty acids are detected in *O. abruptus*. Among them myristic acid, palmitic acid and stearic acid are saturated and rest of the five fatty acids are unsaturated. However, in *O. fuscovittata* a total of six fatty acids were detected. Stearic acid and arachidonic acid are not detected in the latter case. It is noteworthy that both the grasshopper species contained two very important polyunsaturated fatty acids (PUFAs) *i.e.* linoleic acid and linolenic acid. It is also notable that the titers of both of these PUFAs are present in quite high levels.

3.5. Content of Amino Acids in g/100g (per cent) of Protein in Dried Grasshopper Body Tissues

Among the amino acids threonine (Thr), proline (Pro) and tyrosine (Tyr) are present in very high quantity (more than 9 per cent), whereas aspartic acid+asparagine (Asx) and cysteine (Cys)are pretty low (less than 1 per cent).

3.6. Content of Vitamins in mg/100g of Dried Grasshopper Body Tissues

Total five vitamins, namely: retinol, ascorbic acid, thiamine, riboflavin and niacin are perceived in the body tissues of the selected acridids. Among the B vitamins thiamine has been detected in very low amount in both the grasshopper species (less than 0.50 mg/100g). Riboflavin is present in moderate amounts (just above 1mg/100g), whereas retinol, ascorbic acid and niacin content are quite high (more than 4 mg/100g).

3.7. Content of Anti-nutritional Factors in g/100g of Dried Grasshopper Body Tissues

The presence of four anti nutritional factors — oxalate, tannin, phytin, and phytin bound phosphorus (phytin P) are estimated so far for the grasshoppers of interest. All of them are present in very low amount. Among the estimated anti-nutritional factors, tannin is the highest (in between 1g/100g and 2.45g/100g), on the contrary the amount of phytin P is the least.

3.8. Elaboration

Dietary protein is one of the most important constituents in feed of vertebrates, and fish is of no exception. In general, higher protein levels and higher protein to energy (P/E) ratios can produce significantly better growth and feed utilisation in fish. The grasshoppers of our interest reveal that they are high in protein contents which prove this group to be a good protein resource. Protein is composed of mainly twenty amino acids. Among these, total ten are important in fish physiology namely: arginine, histidine, isoleucine, leucine, lysine, methionine, phenylalanine, threonine, tryptophan and valine. The presence of all these amino acids is perceived along with the others in the grasshoppers of our interest. These results are very encouraging because six of the ten essential amino acids (*i.e.* Val, Leu, Arg, Trp, His and Thr) are present in more amounts than the usual need of the fishes. However, despite having a good amount of amino acids, sometimes the edible insects are deficient in some. As for example studies on other edible insects like Mormon cricket (*Anabrus simplex*) and house cricket (*Acheta domesticus*) exhibits these insects to be methionine deficient. Similarly, the essential amino acids of six lepidopteran species are also found to be lacking in methionine. Amino acid profile of *O. fuscovittata* and *O. abruptus* resembled the above stated reports as the content of methionine, cysteine and lysine are below 3 per cent in these insects. Therefore we support the conclusion of a group of scientist in China that insects might be unsatisfactory as the only source of dietary protein due to limiting amino acids, but would be extremely beneficial as a supplement. Fat is one of the major nutrients for any living organism. Insects usually have higher fat content at the larva and pupa stages, while adults have relatively lower amount. Among grasshoppers the fat content of adult Chinese grasshopper *Oxya chinensis*is found to have only 2.2 per cent, whereas the same of *Sphenarium purpuracens* is about 5.75 per cent. A lower fat content (nearly 5-7 per cent) is also observed in the grasshoppers. Although the fat content of most of the edible insects is lower than the optimal need of 10-20 per cent for fish, still they could play an important role as a supplementary diet. Unlike other animal fat, edible insects have a higher essential fatty acid content which is necessary for animal nutrition, and hence edible insect fat has a high nutritive value. The fatty acid contents of our grasshoppers when compared to that of the variegated grasshopper *Zonocerus variegates* it is found that *O. abruptus* has a total of eight fatty acids, while *O. fuscovittata* has six. On the other hand eight fatty acids have also been identified in the Chinese grasshopper *A. cinerea* where palmitotelic and lauric acid have been detected instead of arachidonic and eicosenoic acid. Although it is told that edible insects are rich in protein and fat, they are not so rich in carbohydrate, though it differs from species to species usually ranging from 1-10 per cent. Nevertheless, larva of *Cirina forda* which is relished as a human food in Nigeria contains near about 38 per cent of carbohydrate. Edible insects have been found to contain good amount of minerals. Among them the report of sodium, potassium, calcium, zinc, iron and magnesium predominate the literature. Another report from Mexico said that edible insects under the orders Orthoptera, Lepidoptera and Hymenoptera have much less variation in mineral contents. The report continued stating that these insects are mostly low in Na, and sometimes high in Ca, Zn, Fe, K and Mg. *O. fuscovittata* and *O. abruptus* showed a similar trend as Mg is present in the highest amount (more than 6mg/kg) followed by Ca and Zn. But we cannot detect Na and K, though a small trace of Mn (less than 1 per cent) is there. Reports are prevalent that show the edible insects to be rich in vitamin contents too. As for example rich vitamin content in some edible caterpillars of Attacidae family is reported from Zaire. According to some workers edible insects are rich in vitamin B group such as thiamine, riboflavin and niacin. The presence of these three vitamins is also observed in case of our selected grasshopper species where niacin is present in high amount but rest of the two are a little lower. In addition to this, two other vitamins *i.e.* vitamin C and vitamin A are also present in quite high amount.

Anti-nutritional factors are usually present in plant materials. However, many phytophagous insects have been identified to retain these materials in quite a good amount. Hence it is recommended to analyse these anti-nutrients if a phytophagous insect is considered as food, but this kind of study is not much frequent yet. Among the anti-nutritional factors tannin, oxalate and phytate are detected in the grasshoppers of our interest. A scientific study from the Ondo state of Nigeria reported tannin percentage in grasshoppers to be 1.05g/100g, which is almost similar to the amounts observed in *O. fuscovittata*, but a little higher amount (about 2.5g/100g) is observed in *O. abruptus*. However, this amount is itself quite lower than the ones found in cereal grains and other plant food materials. The percentage of phytate is usually quite high in grasshoppers, but findings in our grasshoppers contradicted with this notion, because radically low amount of phytate is observed in these grasshoppers. However, the oxalate content is a little higher in case of our grasshopper species compared to other edible insects like *Oryctes monoceros* and *Cirina forda*. But this amount is also under the tolerance limit because much higher amounts have been observed in various plant food materials.

3.9. Conclusion

Proximate composition of the grasshoppers reveals that they are nutritious with high protein contents. Amino acids, fatty acids, vitamins and minerals are present in fairly good amount. Though anti-nutritional factors are definitely present, but are in very low amount and within the tolerance limit. Thus it can be safely stated that these insects could be a good alternative source of not only protein, but also important vitamins. This positively indicates towards a serious effort of mass rearing these grasshoppers in controlled conditions as a mini-livestock so that it could supply more and more amount of alternative nutritious food for the formulation of protein rich diet for the ornamental fish industry and thus lowering down the over-exploitation of fish meal.

Chapter 4

Mass Culture Strategies of Grasshoppers

4.1. Introduction

From the previous chapter we have seen that grasshoppers are rich in nutrients and the anti-nutritional factors present are within the tolerance limit. Studies on the biology, ecology and behaviour of short-horn grasshoppers reflected their high reproductive potential and rapid life cycle that leads to a very high annual biomass of these creatures. Because of their ubiquitous presence, grasshoppers already are a potential food for many organisms in nature. Hence, we could harvest these nutritionally rich creatures during their outbreak periods to formulate alternative feed for our livestock. But for a sure and constant supply, their farming is essential. In this context, the feasibility of farming of a grasshopper *Acrida exaltata* (Walker) was assessed. As *A. exaltata* is a bivoltine species, it is unable to produce the biomass needed to feed the livestock because these industries need a constant supply of alternative feed resource, which require production of a huge biomass in permanent acridid farms. A substantial biomass can only be obtained if a multivoltine species can be mass reared successfully. A preliminary work on biomass production of *O. fuscovittata* has been already conducted in order to utilise this natural resource as an alternative feed for poultry and fish, which established that this species is very easy to culture and due to its high reproductive potential and survival rate it has the potential to produce a huge biomass within a year. But to obtain a high amount of biomass of acridids, studies on their growth rate as well as their survival rate are also necessary because rapid growth and less mortality will directly influence their biomass production. The present chapter is focused on estimation of the potential of mass culture of *O. abruptus* and *O. fuscovittata* with four food plants of the Poaceae family (the preferred food plant group for acridids) to determine the effect of these plants on consumption, utilisation, growth and lifespan of the acridids in order to find a suitable plant species that may serve as a major food source in grasshopper farms.

4.2. Obtaining Food Plants to Feed Grasshoppers

Short-horn grasshoppers are known to be graminivorous which means they have a preference to plants under family Gramineae or Poeceae. To feed the grasshoppers four food plants were selected. Soft green leaves of *Cynodon dactylon* (durva grass) were collected from different places to avoid cyanogenic induction which is reported to be enhanced by mechanical cutting time and again from a single site. Seeds of *Triticum aestivum* (wheat),

Figure 4.1: Collection of Adult Acridids of interest in a Grassland Ecosystem.

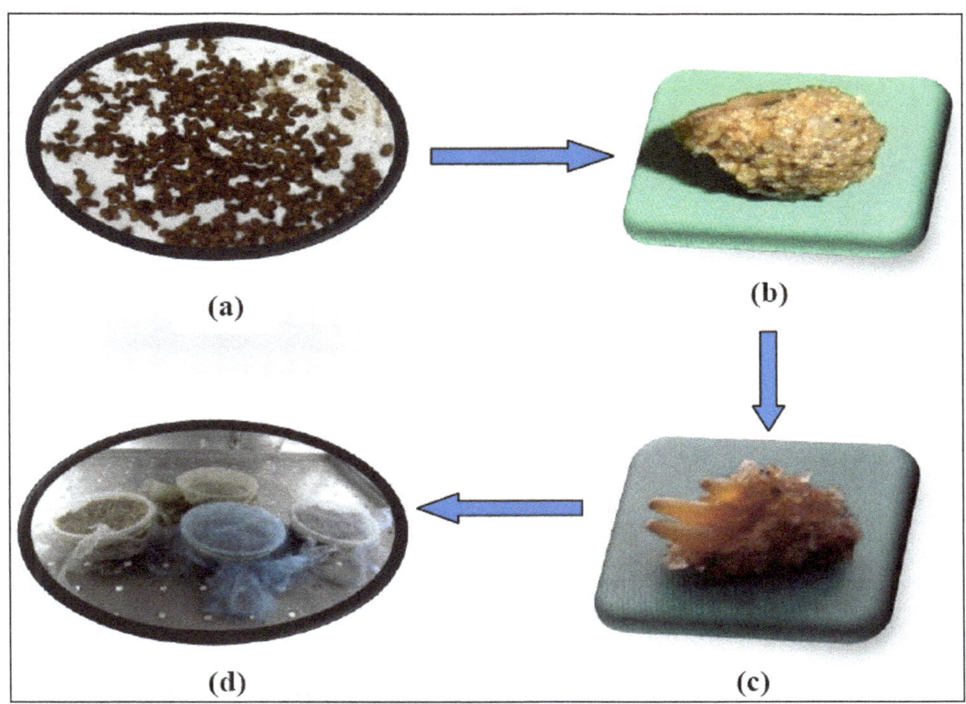

Figure 4.2: The Flow Chart Showing (a) A collection of egg pods, (b) Single egg pod, (c) Arrangement of eggs in a pod, and (d) Egg incubation set up.

Sorghum halepense (Johnson grass) and *Oryza sativa* (rice) were obtained from a local market and grown outdoors. After the seeds had germinated, young leaves were offered to the grasshoppers in glass vials.

4.3. Mass Culture Set up

Adult individuals of *O. fuscovittata* and *O. abruptus* were collected by sampling with standard aerial insect nets and kept in mesh-fitted cages with wooden frame. A tray of sand was kept in the cages and water was sprinkled everyday in order to keep the sand moist so that proper humidity could be maintained. After the insects copulated they laid their pods into the sands. When the young ones hatched out they were kept in transparent plastic jars measuring (11 11 26) cm^3. Then the selected four species of food plants were offered with replicates, having single type of plant species per jar. The floor of the jars was covered with 5cm thick sterilised sand and some water was

Figure 4.3: ***Triticum aestivum*** (L.).

Figure 4.4: ***Sorghum halepense*** (L.) Pers.

Figure 4.5: ***Oryza sativa*** (L.).

Figure 4.6: ***Cynodon dactylon*** (L.) Pers.

Figure 4.7: Acridid Mass Culture in Rearing Jars.

sprinkled daily. The top of the jars were covered with small piece of nylon net.

4.4. Effect of Various Food Plants on Nymphal Growth

Three growth parameters were considered to estimate the growth of the two grasshopper species. These are average daily growth (ADG), growth rate (GR) and specific growth rate (SGR). Among these three indices SGR is considered to be the most efficient one as log value is used for the calculation. The pattern of growth in the two species is presented below (Refer to glossary for details of these parameters).

a) *O. fuscovittata*

The results clearly showed that the individuals of both sexes fed on *S. halepense* grew better in comparison with the other three food plant fed sets as all the three parameters were significantly highest for this food plant fed sets. On the contrary least growth was observed in *O. sativa* fed individuals.

b) *O. abruptus*

Unlike *O. fuscovittata,* the second grasshopper species *i.e., O. abruptus* showed a different trend where highest value of growth parameters were obtained in case of *T. aestivum* fed sets. *S. halepense* fed sets showed a little lower values followed by *C. dactylon* fed ones. However, here also significantly lowest values were obtained in case of *O. sativa* fed individuals.

4.5. Effect of Various Food Plants on Consumption Index (CI) of Adult Male, Adult Female and Nymphs

Consumption index is a measure of food consumed by an individual per day (Refer to glossary). It is a very important measure if we want to comment on most preferred food when choice is given to an individual. Results shown by *O. fuscovittata* and *O. abruptus* are depicted below.

a) *O. fuscovittata*

In case of males, *C. dactylon* and *O. sativa* fed sets showed relatively higher values of consumption index (CI). Though any statistically significant difference was not observed between these two, the mean value was higher in *C. dactylon* fed sets. Considerably low value of male CI was obtained in *S. halepense* fed groups. In females and nymphs also *S. halepense* fed groups showed the lowest values of CI, whereas remarkably high values were obtained in *C. dactylon* fed sets, followed by *T. aestivum* and *O. sativa* in both females and nymphs.

b) *O. abruptus*

The CI of *O. abruptus* showed insignificant variation between food-plants. Nevertheless, the CI value of *T. aestivum* and *S. halepense* fed females was significantly lower than *O. sativa* fed sets, and *T. aestivum* fed nymphal CI was significantly lower than *S. halepense* and *C. dactylon* fed ones.

4.6. Effect of Various Food Plants on per cent Approximate Digestibility (AD) of Adult Male, Adult Female and Nymphs

Approximate digestibility is a measure of how a food is utilised by an individual (Refer to glossary). Following is the description of how the two grasshopper species digest the selected food plants.

a) *O. fuscovittata*

In case of adult males and females *S. halepense* fed groups showed significantly lowest AD values. For females the highest value was obtained in *C. dactylon* fed groups; but for males significantly higher values were observed in both *C. dactylon* and *O. sativa* fed sets where the AD values of these two food plant fed groups did not vary significantly. On the contrary for the nymphs highest value of AD was noted in *T. aestivum* fed sets, followed by *C. dactylon* and *S.halepense*, whereas the lowest value was in *O. sativa* fed groups.

b) *O. abruptus*

In this acridid species AD values of both the adult males and females were highest in *S. halepense*, and lowest in *O. sativa* fed sets. Nymphal AD was also highest when fed on *S. halepense*, followed by *T. aestivum* and *O. sativa*; and the lowest value was found in the insect group fed with *C. dactylon*.

4.7. Consumption and Utilisation of Food Plants by Male and Female Acridids in Nymphal Stage

In the nymphal stage the utilisation of food materials could be indicated by two vital indices, ECI (Efficiency of conversion of ingested food to body substance) and ECD (Efficiency of digested food to body substance). We are not going to the methodology here, about how these two are measured. A detailed description regarding them is discussed in the glossary section. We will now concentrate on the results below.

a) *O. fuscovittata*

The pattern of consumption and utilisation in terms of ECI and ECD of males and females of *O. fuscovittata* followed a similar path as seen in case of growth parameters. Here also for both the sexes *S. halepense* fed groups showed the highest values followed by *C. dactylon*; whereas *O. sativa* fed sets had the least values of ECI and ECD.

b) *O. adruptus*

Highest values of ECI and ECD of *O. abruptus* were obtained when the sets were fed on *T. aestivum*. ECD of both sexes and ECI of males did not show any significant variation in rest of the three food plants; however, for females second best ECI values were obtained in *S. halepense* fed groups.

4.8. Nymphal Duration of the Chosen Acridids Fed with Various Food Plants

Whenever we are keen to know how growth is affected by a food, it is very important to know how much time it takes to become adult when an insect is being fed by a certain food. The nymphal duration, *i.e.* the time lapse from newly hatched nymph to the adult form is described in the two grasshopper species of our interest is given below when they were fed on various food plants.

a) *O. fuscovittata*

Nymphs took remarkably short time to mature when fed on *S. halepense*. Insignificant results were obtained between *C. dactylon* and *T. aestivum* fed individuals, whereas highest nymphal duration was observed in *O. sativa* fed groups.

b) *O. abruptus*

Here the highest nymphal duration was obtained in the sets fed on *O. sativa* and *T. aestivum* fed sets had a significantly lower value than the ones fed on *C. dactylon*.

4.9. Adult Male and Female Life Span of Selected Acridids Fed with Various Food Plants

Adult life span is a very important aspect to be considered because unless and until the adult individuals will live for a certain amount of time they won't be able to copulate and lay eggs. Result of the effect of food plants on adult life span of *O. fuscovittata* and *O. abruptus* is portrayed below.

a) *O. fuscovittata*

Longest life span of adult males was observed when they were fed on *C. dactylon* and *S. halepense*. Life spans of *O. sativa* fed adult males were extremely short. Although life span of adult females gave a similar trend as in the males, females always lived longer than their male counterparts.

b) *O. abruptus*

Adult male life span of *O. abruptus* was higher when the insects were fed on *C. dactylon* and *T. aestivum*. On the other hand *S. halepense* fed sets were significantly lower than *C. dcatylon* and *T. aestivum* fed ones.

4.10. Elaboration

Grasshopper feeding pattern and growth response varies with crop plants and grass species that they consume. Various reports say that in grasses there are a number of factors like secondary chemicals, physical deterrents and nutritional value that can influence grasshoppers at different stages of development. The experiments involving rearing of *O. fuscovittata,* and *O. abruptus* clearly supports the above statement. Feeding on *S. halepense* gave the shortest nymphal duration for *O. fuscovittata,* while feeding on *T. aestivum* gave the shortest result for *O. abruptus*. In a similar work reported from Nigerian grasshopper *Oedaleus nigeriensis* it has been reported that nymphal duration was significantly low when they were fed on *Seteria gracilipes* and *Axonopus compressus.* In *Oxya nitidula* the shortest nymphal periods were obtained when rearing was on *Panicum maximum.* Adequate diet is expected to reduce nymphal developmental period. Proper nymphal development is regarded as the most crucial part for regulation of a grasshopper population. So to ensure adequate nutritional requirements, in nature, grasshoppers may favour habitats with mixed food plants. Studies on the nutritional ecology of many grasshopper species have revealed that mixed food plants are better than any single type. For example, mixed plant diets are superior for *Melanoplus bivittatus,* that promotes higher survival, larger adults and higher growth indices than any single plant diet. It was also evident that individual polyphagous grasshoppers do switch between dietary items and intake mixed food more than the two foods offered separately. Regardless the diet is a single type or a mixed one, grasshopper body size influences food consumption. Heavier insects consume more food may be because of higher metabolism. In the two selected grasshoppers of our interest the females are bigger and heavier than males like other Orthoptera species, and hence it has been observed that in most cases females consumed more, or have a greater efficiency of conversion of ingested food than the males of their respective species. Growth parameters (GR, per cent ADG and per cent SGR) and food utilisation (ECI and ECD) were highest in *T. aestivum* and *S. halepense* fed groups respectively for *O. abruptus*. In case of *O. fuscovittata* individuals fed with *S. halepense* showed highest results for all the growth parameters and ECI and ECD. On the other hand *O. sativa* fed groups of both the species almost always showed minimum growth and food utilisation. *O. abruptus* consumed *O. sativa* maximum at the adult stage, whereas *O. fuscovittata* fed least on this plant. These results support the view that dietary imbalance often alters feeding behavior to compensate for suboptimal meals. From the results of food consumption, utilisation and growth it can be concluded that a grasshopper that encounters plants low in critically needed substance (proteins for example) may reject these food plants (as shown by *O. fuscovittata*) and choose another. If choice is not available they may consume greater amount of the lower graded food plants (as shown by *O. abruptus*) to compensate for the lower nutritional value. Moreover, if there is a food plant present with adequate nutrients they may consume these plants in a lower amount (as shown by *O. fuscovittata* for *S. halepense*) to get the optimal nutrition. Such results were also obtained on some other grasshoppers like *Chrotogonus lugubris, Schistocerca gregaria, Locusta migratoria, Zonocerus variegates, Heteracris littoralis* etc. The unsatisfactory results of *O. sativa* fed groups may seem ambiguous because these acridids are usually found near agricultural fields and are considered as a pest of rice. In this context one may justify that females of acridids usually lay eggs under moist soil near swamps, ponds and other natural water bodies and newly hatched nymphs feed on soft green grass (and not rice seedlings). Maybe these insects consume rice leaves only after attaining adulthood. As the individuals in the experimental sets were offered *O. sativa* from the very first day the nutrition might be inadequate that led to growth retardation, late maturation and very short adult life span.

4.11. Conclusion

We can conclude the present chapter by saying that *S. halepense* is the most suitable food plant for the nymphal stages of *O. fuscovittata* and *T. aestivum* and *S. halepense* are the two most suitable food plants for *O. abruptus* nymphs. These results prove that food plant is a very important variable as it can exhibit so many effects on growth, food consumption, utilisation and lifespan of a grasshopper. In the present case we have evaluated the effects of food plants as the only type of food provided per set. But these food plants may also be an excellent supplemental food as part of a mixed diet and hence further testing should focus on the use of a mixed diet and their consumption and utilisation by the insects in order to obtain a better picture.

Chapter 5

Annual Biomass Production of Selected Grasshopper Species

5.1. Introduction

In chapter 2 our main focus was on growth, food consumption, utilisation and life span of the grasshoppers fed on various food plants. But to finally conclude on which plant is most suitable for them for biomass production, assessment of reproductive potential and survival of nymphs is a must. Moreover to conclude on which grasshopper species is better suited for establishing "grasshopper farms as mini-livestock", one should make an attempt to estimate which one could produce higher biomass annually. In this context the aim of the present chapter is first to determine the most suitable food plant for the chosen species and then to determine how much biomass they could produce annually. It is obvious that more the biomass, more efficiently the grasshopper farms will be able to supply their insects to the companies in order to manufacture supplementary feed for livestock industries.

5.2. Estimation of Grasshopper Biomass

Nymphal survival percentage, number of egg pods laid per female, number of eggs hatched per pod, sex ratio and dry body weight of adult individuals were estimated for the chosen grasshopper species *i.e., O. fuscovittata*, and *O. abruptus* fed with *S. halepense*, *T. aestivum*, *O. sativa* and *C. dactylon*. In the present case these estimations have been conducted with the same set up used in the mass culture experiment. The food plant with best results for each of the three acridid species is considered for the estimation of their yearly biomass production. Energy of each of the selected acridid body tissue is estimated in terms of kJ/100g by oxygen bomb calorimeter. Energy of the body tissues of males and females of respective species did not show significant variations; hence I have taken a mean value of energy to calculate the biomass in terms of energy. For calculation of annual biomass total number of individuals to be produced in a year has been first estimated. This is followed by multiplying the numbers with individual dry body weights of respective species to obtain the total dry biomass. Likewise, total biomass produced in terms of energy is again calculated by multiplication.

O. fuscovittata shows best results when fed on *S. halepense* and *C. dactylon* for the parameters of fecundity, fertility and survival. On the other hand fecundity, fertility and survival of *O. abruptus* shows the best results when fed with *C. dactylon* only. Here we should recall the results of the previous chapter where food consumption, utilisation and growth parameters showed favorable results for *S.halepense* fed *O. fuscovittata* and *T. aestivum* fed *O. abruptus*. It could also be kept in mind that *S. halepense* fed *O. abruptus* almost always gave the second best results. Combining these with that of fecundity, fertility and nymphal survival in the present chapter, *S. halepense* could be identified as the favourable food plant for calculation of annual biomass production of *O. fuscovittata*. On the contrary results were a little bit critical for *O. abruptus*. In this species *T. aestivum* fed sets gave best results for nymphal growth, food consumption and utilisation followed by *S. halepense* fed ones. But *T. aestivum* fed groups showed a very low nymphal survival (about 31.11 per cent), hence the second best food plant *i.e. S. halepense* could be selected as the suitable plant species for nymphal rearing of *O.abruptus*, as in this case relatively higher survival rate is obtained. On the other hand, because of a greater adult life span (ref. chapter 2), fecundity and fertility, *C. dactylon* is selected as the suitable food plant for the adults during the calculation of yearly biomass production of these grasshoppers.

In the course of calculation of probable annual biomass, sex ratio and energy of the grasshopper body tissues is also estimated in addition to nymphal survival, fecundity and fertility. Then the calculation of total yearly biomass that could be produced in terms of numbers is carried out. As *O. fuscovittata* is a tetra voltine species they have the ability to complete four life cycles annually. Hence, their biomass is calculated for four generations, whereas *O. abruptus* being a trivoltine species its biomass is calculated up to three generations starting with a single male and female in each case. According to current estimation at the end of a year *O. abruptus* can produce only about 348 males and 436 females (*i.e.* total 784 individuals). But the results of *O. fuscovittata* is extremely fascinating which shows this insect has an ability to produce a huge biomass of about 2,90,389 males and 4,35,583 females (*i.e.* total 7,25,972individuals) in a year from just one pair. Annual biomass produced in terms of dry weight is estimated to be about 13.6g of males and 22.6g of females (*i.e.* about 36.2g total dry biomass) for *O. abruptus* and about 13.4 kg of males and 28.3 kg of females (*i.e.* about 41.7 kg total dry biomass) for *O. fuscovittata*. Estimation of yearly biomass production in terms of energy also shows *O. fuscovittata* to have the highest value of near about 2,59,927 kJ for males and 5,50,929 kJ for females.

5.3. Elaboration

Biomass is directly dependant on fecundity, fertility and nymphal survival which are significantly varied by the food plants on which the grasshopper feeds. In other words it can be said that food plants do have significant effect on biomass production. The results of variations of nymphal survival are in concert with the differential mortality in a polyphagous grasshopper *Parapodisma subastris*, when they were reared on different plant diets. Long back in 1966, Sir Boris P. Uvarov, the all-time great scientist in the field of grasshopper research, reported that fecundity appears to be dependent on the relative value of a food plant both for the survival of a female and for the reproductive activity. It is also observed that grasses of different quality and quantity affect the egg production in locusts (*i.e.*, migratory grasshoppers)where it is assumed that feeding may affect the development of fat body during somatic growth, which later affects vitellogenesis and egg development of insects.

Host plant quality also affects insect reproductive strategies: egg size and quality, allocation of resources to eggs, and choice of oviposition sites may all be influenced by plant quality. It is also believed that host plant quality may significantly affect offspring quality as well as quantity. On a poor-

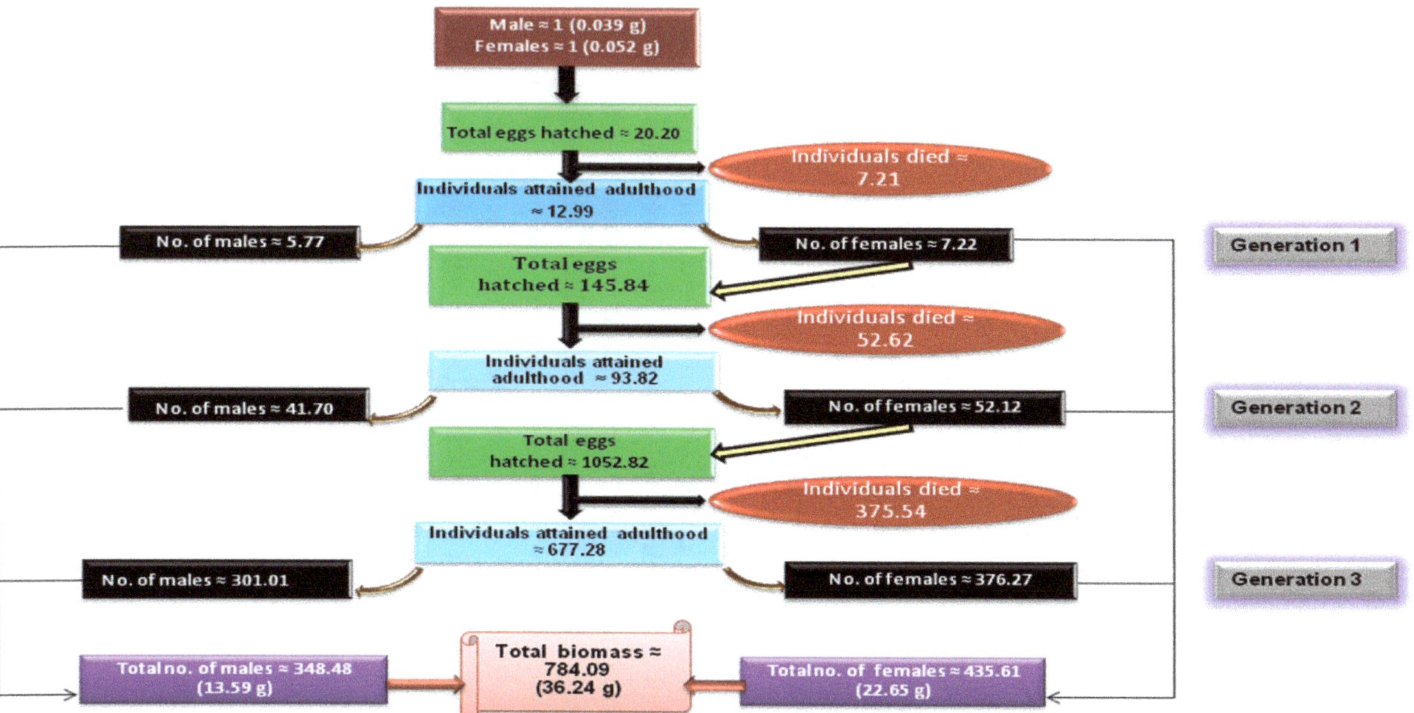

Figure 5.1: Projected Annual Biomass Production of *O. abruptus* in Terms of Number and Dry Weight when Fed on Suitable Food Plant.

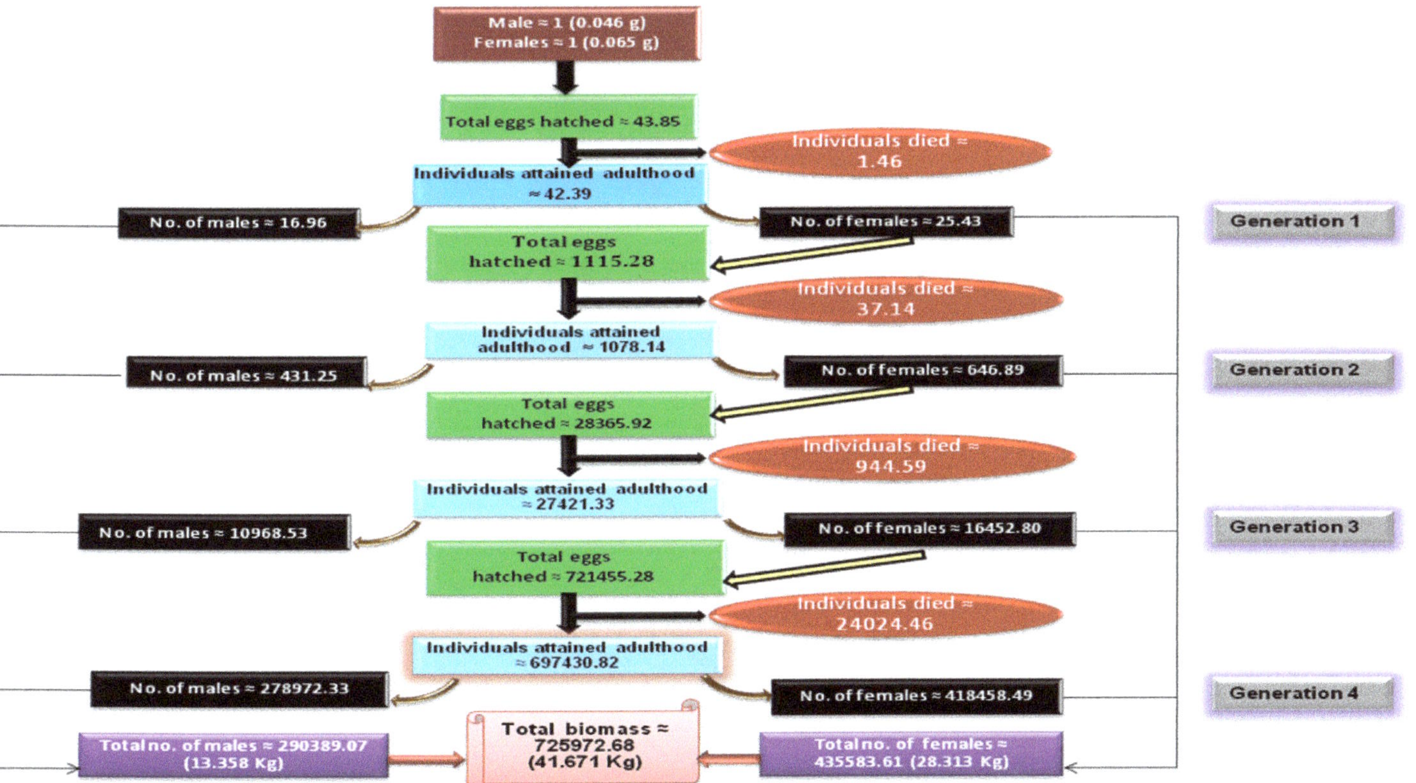

Figure 5.2: Projected Annual Biomass Production of *O. fuscovittata* in Terms of Number and Dry Weight when Fed on Suitable Food Plant.

quality host plant a female insect may either lay few good-quality eggs or a large number of poor-quality eggs. Moreover, the host plant quality is also assumed to affect the male insect's contribution to the production of fertile eggs. In grasses there are a number of factors that can influence grasshoppers at different stages of development, including secondary chemicals, physical deterrents and nutritional value.

5.4. Conclusion

The results of the present chapter on the selected grasshopper species clearly reveals that fecundity, fertility and nymphal survival are greatly affected by the selection of food plants, resulting in a definite variation in biomass. It is observed that amongst the two, *O. fuscovittata* has the capability of producing the highest biomass. Thus the huge biomass of this nutritionally rich species can be utilised as a livestock feed. These results support the view of establishing "acridid farms as mini-livestock" where *O. fuscovittata* could be mass reared using *S. halepense* as their food plant in controlled conditions with the system to prevent their outbreak. Ultimately this huge biomass might be supplied to the feed developing companies to make alternative low cost protein rich diets for the livestock.

Chapter 6

Formulation of Various Protein Rich Diets Using Grasshoppers as a Supplement

6.1. Introduction

From the results of the previous chapters it has been found that both the chosen acridid species are nutritionally rich, and among them *Oxya fuscovittata* could yield a very high biomass if it is produced in mass scale in grasshopper farms with its suitable food plant *Sorghum halepense*. We have also come to a conclusion that they could be used as alternative protein supplement for livestock. Among the livestock the present book focuses on the ornamental fisheries industry and for this purpose a model fish is needed which could be easily reared, could complete life cycle within a short period and easy to breed so that their growth, life span and reproductive potential could be measured. In this book we will focus on a live bearer fish, *Poecilia sphenops* (Valenciennes) or more commonly "black molly" for such study as it is one of the most common ornamental fishes to be seen in the aquaria.

6.2. *Poecilia sphenops* (Valenciennes): A Brief Description

Black mollies are the result of crossing a black form of *Poecilia latipinna* with *Poecilia sphenops* and are themselves known as *Poecilia sphenops* (Valenciennes). They may grow up to 6.0 cm with an average length of 3.5 - 4.0 cm. Males could be easily distinguished from females with their characteristic gonopodium, which is a modified anal fin used to inseminate females. They prefer hard fresh water to brackish water with a pH range of 7.5-8.2. They need a temperature range within 18 C - 28 C. Like other live bearers, mollies are true omnivores, they feed on worms, crustaceans, insects and plant matter.

Distribution

These are the native species of Central and South America (including Mexico to Colombia).

Systematic Position

Phylum – Chordata

Class – Teleostomi

Subclass – Actinopterygii

Order – Cyprinodontiformes

Family – Poeciliidae

Genus – *Poecilia*

Species - *Poecilia sphenops*

So, *O. fuscovittata* is a suitable acridid that could be used as a protein supplement, and *P. sphenops* is a suitable model ornamental fish to work on. But before beginning a feeding trial, it is first needed to formulate appropriate feed for this fish where the conventional fish meal will be replaced by *Oxya* meal.

6.3. Feed Formulation Strategies

For the formulation of various supplementary diets *Oxya* meal, fish meal, mustard oil cake (MOC) and rice husk were used as major ingredients. Carboxymethylcellulose (CMC) was used as a binder for the diets and calcium, table salt (NaCl) and vitamin-mineral mixture were used as additive. First of all proximate composition of the major ingredients were

Male

Female

Figure 6.1: ***Poecilia sphenops*** (Valenciennes).

carried out as discussed in chapter 1. Then five diets were prepared gradually replacing fish meal by *Oxya* meal with the ratio of 100:0, 75:25, 50:50, 25:75 and 0:100. The diet having "fish meal: *Oxya* meal" with the ratio of 100:0 was considered as control because there was no supplement of *Oxya* meal. The other diets were named D1, D2, D3 and D4 respectively. The amount of ingredients of all of the formulated diets were mixed in such a way so that each of them contain about 40 per cent crude protein because according to literature mollies need this much protein in their diets for optimal growth. Maintaining the amount of protein by adjusting the proportion of ingredients could be done by the Pearson's square method. The method has been elaborated taking an example below:

6.4. Formulation of a Supplementary Diet where Fish Meal and Oxya Meal is Mixed in 50:50 Ratios

In the present book I have made an attempt to discuss about formulation of diets having 40 per cent crude protein using *Oxya* meal, fish meal, mustard oil cake (MOC) and rice husk as major ingredients, that could be fed to the black mollies. Crude proteins of these ingredients were recorded 48.31 per cent for fish meal, 64.07 per cent for *Oxya* meal, 5.44 per cent for rice husk and 30.63 per cent for MOC. Now, to prepare a diet where fish meal and *Oxya* meal are mixed in 50:50 ratios, at first one should decide the inclusion level of fish meal and/or *Oxya* meal to the formulated diet. Here I have decided an inclusion level of 60 per cent of the total diet for these two ingredients and 5 per cent of the total diet was left for binder and other additives. Next, it is necessary to know the amount of protein that will be contributed by fish meal and *Oxya* meal. It is apparent that in 60 per cent inclusion level the amount of fish meal and *Oxya* meal mixed in 50:50 ratios will be 30 per cent of the total diet each. Hence 30 per cent fish meal will contribute 14.50 per cent (48.31 30/100) protein and 30 per cent *Oxya* meal will contribute 19.22 per cent (64.07 30/100) protein; or, total 33.72 per cent protein will be contributed. Therefore the other 35 per cent of the formulated diet (60 per cent + 5 per cent is already utilised) will have to make up 6.28 per cent of protein by rice husk and MOC to formulate a diet having 40 per cent crude protein. If this portion of the formulation is treated separately, a mixture of rice husk and MOC must contain 6.28 100/35 = 17.943 per cent of protein. The appropriate proportion of rice husk and MOC needed in order to provide 6.28 per cent of protein can be calculated by constructing a "square" as follows:

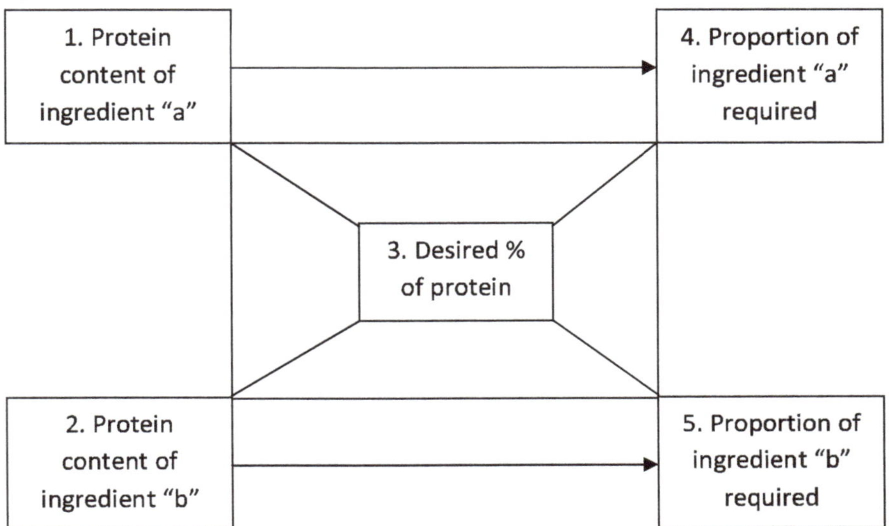

Substituting the values of rice husk and MOC and subtracting 3 from 1 and 3 from 2 (ignoring negative signs) we get the values of 5 and 4 respectively:

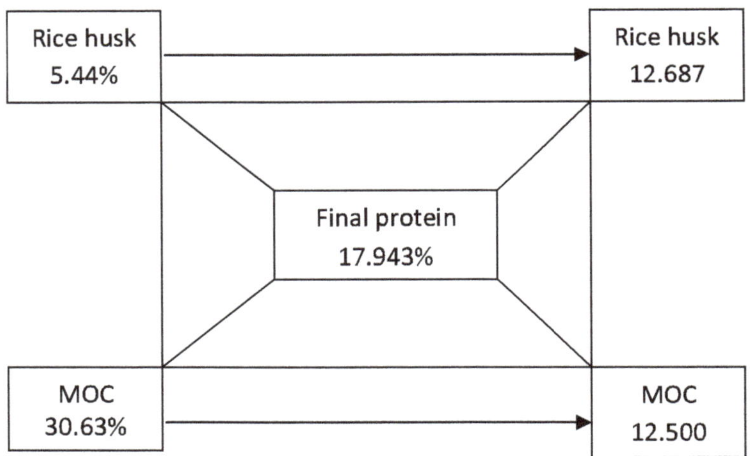

The amount of ingredient to be used is:

$$Rice\ husk = \frac{12.687}{12.687 + 12.500} \times 100 = 50.37\%$$

$$Rice\ husk = \frac{12.687}{12.687 + 12.500} \times 100 = 50.37\%$$

These ingredients however, should constitute only 35 per cent of the total formulated diet. Therefore the actual proportion of rice husk and MOC to be incorporated would be 0.5037   35 = 17.63 per cent and 0.4963   35 = 17.37 per cent of the final diet respectively.

6.5. Formulated Diets

Figure 6.2 shows the major ingredients *i.e.* rice husk, *Oxya* meal, MOC and fish meal. Estimation of proximate composition of the ingredients revealed that crude protein level was highest in *Oxya* meal (more than 64 per cent). Crude fat content of fish meal was found to be highest that was more than 9 per cent. Percentage of carbohydrate, total ash and crude fiber showed higher values for rice husk. MOC showed the highest value only for nitrogen free extract (NFE).

Figure 6.2: Major Ingredients of the Formulated Diets.

After getting the proximate composition of the major ingredients their amounts needed in the formulated diets were calculated by Pearson's square method as already described in the previous section. After formulation, diets were first made to a dough form, and thereafter globule forms were made ready for the feeding trial as shown in Figures 6.3 and 6.4.

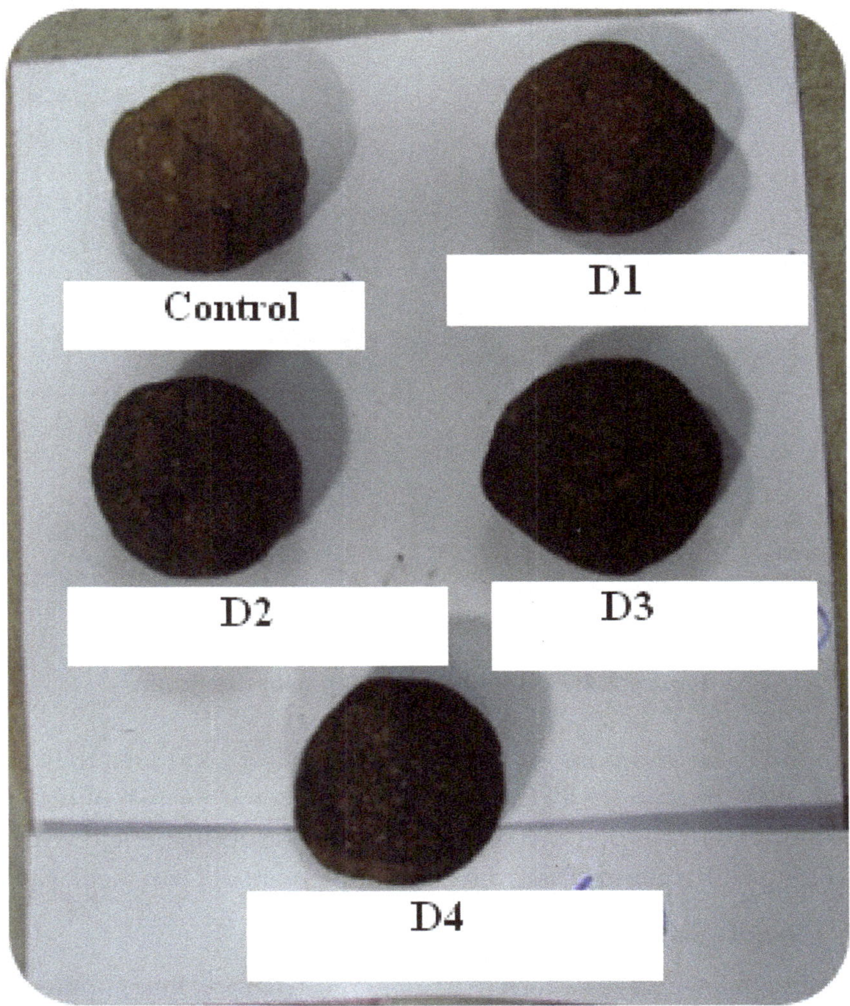

Figure 6.3: Formulated Diets in Dough Form.

6.6. Elaboration

Protein is one of the most important constituents in feed which is important for fish growth. Before designing any experiment on the effect of formulated diets on growth of fishes, the dietary percentage of protein needed for their optimal growth should be obtained first. Previous works on poeciliids like *Poecilia reticulata* reported that 30-40 per cent of protein in feed is optimum for their growth. 40 per cent dietary protein content is

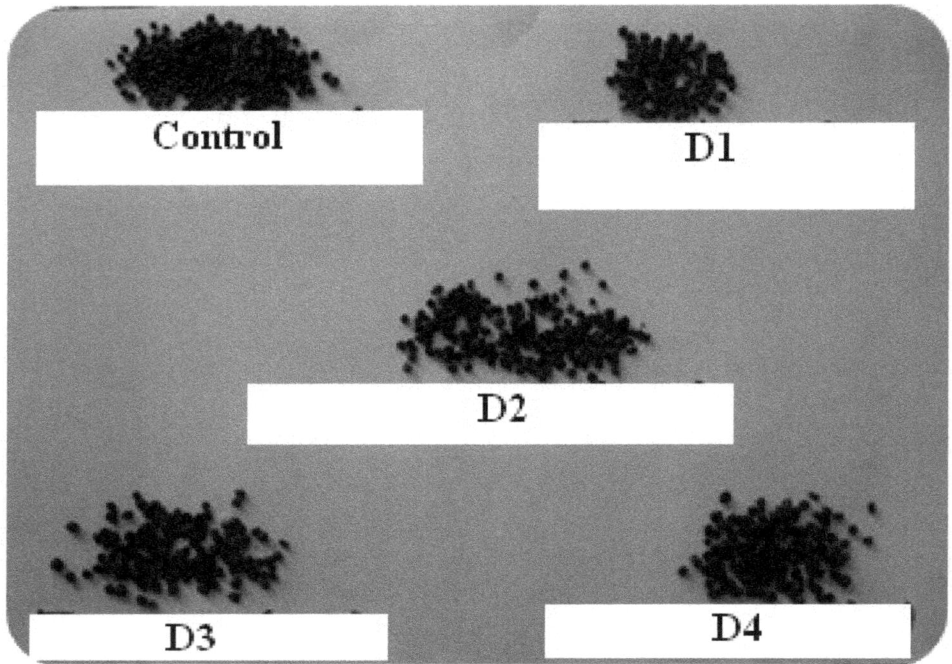

Figure 6.4: Formulated Diets in Globule Form.

also found to be optimum for mollies. Hence it is a good idea to fix 40 per cent of crude protein for all the formulated diets for our fish of choice (*i.e.* black molly). I have tried to formulate four diets supplemented with *Oxya* meal along with a control diet where no supplement of *Oxya* meal was used.

6.7. Conclusion

In this chapter we have seen that one control diet and four supplemented diets were formulated for the feeding trial. However, before that we should be assured of the quality of these diets to conclude whether they are suitable for feeding the mollies at all.

Chapter 7

Quality Assessment of the Formulated Diets

7.1. Introduction

In the previous chapter we have gone through the formulation procedure of one control diet and four other diets supplemented with *Oxya* meal where the crude protein level was decided to be fixed at 40 per cent. We have also come to a conclusion that before feeding trial it is necessary to know the quality of those formulated diets. In this context the aim of the present chapter is to deal with the diet quality assessment to conclude on their suitability as feed for fishes. For this assessment we will now focus on the variables like proximate composition of the formulated diets, suspension time of those diets when put in water, stability of the diets in water and their acceptability as food by the chosen ornamental fishes.

7.2. Proximate Composition and Energy

Crude protein, crude fat (ether extract), carbohydrate, total ash (TA) and nitrogen free extract (NFE) of the diets were first estimated. Energy contents of the diets were estimated by oxygen bomb calorimeter and expressed in kJ/100g. Protein to energy (P/E) ratio is an important measure for the quality assessment of the formulated diets and in the present case it is expressed as "mg of protein/ kJ of energy".

7.3. Estimation of the Suspension Time of the Formulated Diets

Suspension time is the time taken by the formulated diets on water surface before sinking. This could be recorded by a stop watch. For the present case the feed pellets were put in 100 ml water filled beaker and their suspension time was then recorded.

7.4. Water Stability of the Formulated Diets

Water stability was measured by the following method: 2.5 grams of oven dried pellets of each diet were taken in triplicates. They were put in a piece of nylon cloth separately and knotted with a piece of string to make small bundles. These were suspended with a string on a stand and then were immersed in a 250 ml beaker filled with 150 ml of water. The systems were left for about 1 hour. After the set time of 1 hour, the bundles were removed and allowed to drain under tap water for 5 minutes. Thereafter they were oven dried overnight and weighed to get the final dry weight. Water stability was then calculated by the given formula:

$$\%\ \text{water stability} = \frac{\text{Final dry weight of feed}}{\text{Initial dry weight of feed}} \times 100$$

7.5. Acceptability Test of the Formulated Diets by the Selected Ornamental Fish

Feed acceptability was assessed as the time taken by the first fish to strike the first particle after the diet was dropped into the tanks. The time was recorded in seconds using a stop watch. Then the feed acceptability was estimated as the reciprocal of striking time using the following formula:

$$\text{Feed acceptability} = \frac{1}{\text{Feed striking time (sec)}}$$

7.6. Proximate Composition and Energy Contents of the Formulated Diets

Almost 40 per cent of crude protein level was obtained in each of the diets. Similarly the values of crude fat (ether extract) and carbohydrate contents were also very similar between the formulated diets; however, in the control the value was recorded slightly lower than the others. Total ash had a result which was reciprocal to the values of carbohydrate where the control diet showed higher values. There was not much variation in the nitrogen free extract (NFE) of control, D1 and D2 diets, but the values of D3 and D4 were comparatively lower. Similarly in case of crude fiber (CF) the values of control, D1 and D2 diets again did not vary significantly, however they were comparatively higher in D3 and D4. Energy values were quite similar in all the diets, but the control and D4 had significantly lowest and highest mean values respectively. Protein to energy ratios (P/E) were almost the same (nearly 21mg/kJ) for all the diets.

7.7. Suspension Time, Water Stability and Acceptability of the Formulated Diets

The results of suspension time, water stability and acceptability of the formulated diets as a food by the chosen fish *Poecilia sphenops* were estimated for the present case where the values did not have much variation diet to diet. However, D3 had a significantly lower suspension time than D1 and D4. Mean values of water stability were more than 95 per cent for each diet and no significant variation was obtained between the diets. Results of the acceptability test revealed that the time taken by the first fish to strike the

Figure 7.1: Globules of a Formulated Diet Floating on Water Surface.

Figure 7.2: One Black Molly Striking a Globule of a Formulated Diet to Feed.

first particle of the formulated diets was within about 6-7 seconds for each case. Hence, no significant variation was yet again obtained.

7.8. Elaboration

From the results it is evident that five iso-nitrogenous diets were successfully formulated with gradual replacement of fish meal by *Oxya* meal where the crude protein contents were about 40 per cent that fulfills the optimum need of mollies. Along with 40 per cent crude protein level, minimum of 40 per cent NFE is also necessary for these fishes. It is noteworthy that in the present formulation though the control, D1 and D2 diets showed nearly 40 per cent of NFE, D3 and D4 diets had a little lower value (about 34 per cent and 28.5 per cent respectively). It is a well known fact that 10-20 per cent of lipid in fish diets gives optimum growth rates without producing an excessive fatty carcass. In this context it may seem that my diets contain a little lower amount of fat (near about 5-6 per cent). However, there exist some report that support the view that live bearer fishes like mollies and guppies are fed with diets having 3-7 per cent of fat in Singapore. This report also supports the view that the formulated diets of the present study contain sufficient amount of fat for our ornamental fish of interest. It is already recognized that fishes apparently do not have carbohydrate requirement, but it could be successfully metabolized by many of them. Especially, herbivorous fishes have been found to metabolize carbohydrates better than carnivorous species

Warm water fish such as mollies are able to use high carbohydrate levels as an energy source. This phenomenon is advantageous because carbohydrate sources are inexpensive as seen in the present case where all of the formulated diets contained a high amount of carbohydrate. Protein to energy (P/E) ratio is another important measure that gives us a clear picture whether the balance between protein and energy contents is maintained in the diets in proper amount. For achieving maximum growth, the P/E ratio in fish diet should range from 19-22 mg/kJ. These reports are encouraging because in the present case all of the formulated diets had P/E value of nearly 21mg/kJ.

Conventionally fishes are fed for 5-7 minutes twice a day. As the formulated diets showed a suspension time of more than 40 minutes for all instances it is quite clear that this is a sufficient time for fishes to consume these feed materials. In spite of the presence of binder the ability of suspension of the feed ingredients have a direct effect on the water stability of the feed. Though the formulated diets for the present study had the same amount of binder, their major ingredients varied to balance the protein percentage. However, there was no significant variation between the values of water stability. After formulation of the diets and after getting satisfactory results regarding the nutritional value, suspension time and water stability it was necessary to be sure whether the experimental fish recognises the diets as a palatable one. The results of low feed striking time positively indicated about the palatability of all the formulated diets.

7.9. Conclusion

All the five formulated diets had almost similar nutritional value and they fulfill the need of mollies. These were water stable and had sufficient suspension time to feed for the fishes. All of the diets were acceptable to them as a feed. Hence these iso-nitrogenous diets are suitable to conduct feeding experiments with *P. sphenops*.

Chapter 8

Feeding Trial with Black Mollies

8.1. Introduction

Fish meal is a traditional source of dietary protein which is used in fish diets throughout the world. Along with fish meal, other traditional protein sources like, soybean meal, groundnut cake etc, are being competed for by the human population as well. Recently, with a global population boom the increasing demand and uncertain availability of these traditional protein sources resulted in the quest of other protein rich alternatives for inclusion into the fish diets. These include earthworms, snail, mussels, periwinkle, lizard, maggot, frog, plants, insects etc.

Among the animal protein sources, insects play an important role especially in areas where human and livestock populations face chronic protein deficiency. Nutritional evaluation of some edible insects has been investigated in various parts of the world that indicated these creatures contain a large quantity of protein and energy. Moreover, various investigators established that most of the edible insects contain phytic acid as the only prominent anti-nutrient that could be easily detoxified during processing like frying, boiling and roasting.

Previous chapters of the present book has already discussed that *Oxya fuscovittata* could be a suitable alternative insect protein source to be included in the diets of the chosen ornamental fish *i.e.* black mollies. These are nutritionally rich, with a great potential to produce a huge biomass in acridid farms. The diets formulated with supplemented *Oxya* meal have been found to contain optimum amount of nutrients. Moreover, suspension time, water stability and acceptability of the formulated diets by black mollies prove that these are suitable feed for them. With this information in mind the present chapter deals with feeding effect of the formulated diets on *P. sphenops* and determination of the most suitable diet on the basis of this feeding trial.

8.2. Collection of Experimental Species

Twenty females and ten males of black mollies (*P. sphenops*) were bought from the local market. They were reared in the laboratory till the females gave birth to sufficient amount of offspring. The feeding experiment was started with 150 fingerlings of similar initial weight (about 0.01 g). They were kept for 7 days of acclimation period prior to feeding.

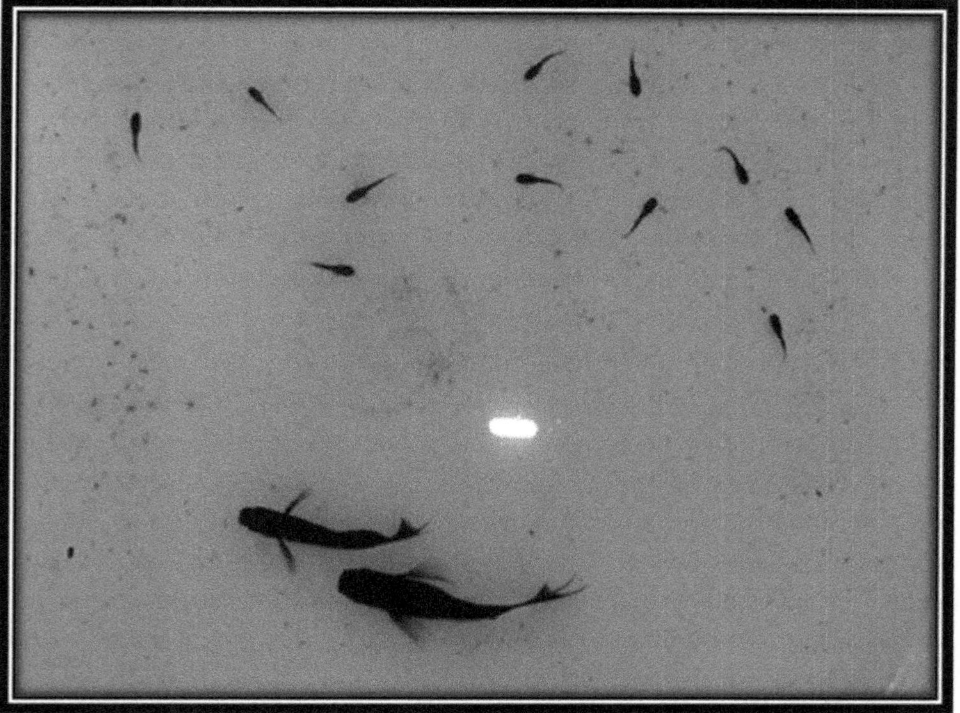

Figure 8.1: The Picture Shows a Pair of *P. sphenops* with their Fingerlings.

8.3. Experimental Setup

150 newly born mollies were divided into 5 sets for testing the effects of one control diet and five *Oxya* supplemented diets on the fishes and placed in experimental tanks. During the feeding trial half of the total volume of water was changed daily from each tank. The water was completely drained from the tanks and thoroughly cleaned with a scrubber once a week. After 176 days the body weight and length of the adult fishes were measured.

8.4. Estimation of Consumption, Utilisation, Survival, Growth and Reproductive Potential of Black Mollies when Fed with the Formulated Diets

Growth was measured in terms of specific growth rate (per cent SGR), average daily growth (per cent ADG), per cent weight gain (PWG) and condition factor (K). Food consumption and utilisation was measured in terms of protein efficiency ratio (PER), feed conversion ratio (FCR), and feed conversion efficiency (FCE).

Survival was measured in terms of percentage of individuals attained sexual maturity, whereas reproductive potential was measured in terms of total number of offspring born per female in the experimental sets.

8.5. Consumption, Utilisation, Survival, Growth and Reproductive Potential of *P. sphenops* when Fed with the Formulated Diets

More than 80 per cent survival was obtained for all the experimental sets without having any significant variation. But for the other estimated indices as a whole there was a tendency of the control diet, along with D1 and D2 fed mollies to have similar results that were significantly better than D3 and D4 fed groups. In case of average daily growth (per cent ADG) and specific growth rate (per cent SGR) there was not much varation between the control, D1 and D2 fed sets, but these values were higher than D3 and D4 fed ones. The values of feed conversion ratio (FCR) also showed better results when fed with the control, D1 and D2 diets. On the contrary unlike ADG, SGR and FCR, the results of protein efficiency ratio (PER) and feed conversion efficiency (FCE) showed marked variation between the control, D1 and D2 fed sets. For PER among these three diets D1 fed groups showed highest values followed by control and D2 fed groups. For FCE, though the values were pretty similar in control and D1 fed groups, the sets fed with D2 showed a slightly lower value. Per cent weight gain (PWG) and condition factor (K) also showed a similar trend, and in both of these cases there was no substantial amount of variation between control, D1 and D2 fed mollies, but then it gradually decreased in D3 and D4 fed sets. Each female of all the experimental tanks gave birth to 13-16 offspring. On an average notable difference of reproductive potential was not observed.

8.6. Elaboration

It is a well known fact that mollies prefer hard water with a pH range of 7.5-8.2, and 18 C - 28 C is the suitable temperature for them. Estimation of water quality during feeding trial of the present fish revealed that the water in the experimental tanks were hard with a temperature and pH within the optimum range. However, dissolved oxygen was a little lower than the optimum need (*i.e.* 5 mg/l) for most of the fishes. The main cause behind this lower DO may be because no aerator was used in the tanks during the feeding period. However, mollies have been reported to get adapted to hypoxic conditions as they could tolerate a level of DO as low as 1mg/l. In this context we could state that the water condition of the experimental tanks were ideal for rearing mollies. Regarding the formulated diets we have already observed that all of them contained the most important components *i.e.* protein, fat and energy in similar amount. However, the values mainly varied in terms of NFE and fiber. The content of protein fat and energy, and the balance of protein and energy (in terms of P/E ratio) were all present in the optimum range for mollies. But along with protein, higher NFE is also important for better growth of mollies. Moreover, fiber content is an essential factor that should be less than 7 per cent in fish diets On the contrary many workers reported that many fishes could grow well after feeding diets containing just above 10 per cent crude fiber; but over 14 per cent fiber level could adversely affect growth. Values of crude fiber in the formulated diets of the present case revealed that though the contents were quite similar in control, D1 and D2, they were a little higher in case of D3 and D4; likewise the NFE contents were similar in control, D1 and D2 diets, whereas a little lower values were obtained in D3 and D4. Apart from only these two criteria all the formulated diets were almost the same. The feeding trial revealed that the results of growth, food consumption and utilisation were very similar when fed on control, D1 and D2. On the contrary the values when fed on D3 and D4 were of a lower grade. Among the very little information in literature on using insects as alternative protein supplement for fish, satisfactory results for *Heterobranchus longifilis* is reported where 50 per cent fish meal was replaced by termite meal. Another work that gave satisfactory results for Japanese carp was in case of 30 per cent cockroach meal that was added to formulate diets. Along with food consumption, utilisation and growth, the measurement of condition factor (K) is also important as it could express the final health of the adult individuals in terms of length and weight at the end of the feeding trial. Results of the condition factor followed a similar trend as observed in case of the other indices already mentioned, because here also the results of control, D1 and D2 diets showed ideal results (more than 1.5). Unsatisfactory results were obtained in case of D3 and D4 fed groups might be because of mollies cannot tolerate a composition where more than 50 per cent of fish meal is replaced by the *Oxya* meal. Or this could be an effect of lower NFE and higher fiber contents of these two diets. In future it is necessary to include higher amount of *Oxya* meal in such a way so that the balance of NFE and fiber is maintained along with protein, fat and energy. Only then one could conclude on the fact that mollies could tolerate more than 50 per cent replacement of fish meal by *Oxya* meal or not.

8.7. Conclusion

Results of this chapter is encouraging because it is quite clear from this study that at least up to 50 per cent of fish meal could be effectively replaced by *Oxya* meal, or in other words it could be stated that fish meal and *Oxya* meal could be added in the formulated diet in 1:1 ratio. This amount of replacement will have no negative effect on the growth and reproduction of black mollies.

Chapter 9

Comparison of Growth Performance of Black Mollies Fed with Formulated Diets and Market Available Fish Diets

9.1. Introduction

Results of the feeding trial of the fish *Poecilia sphenops* revealed that among the formulated *Oxya* meal added diets, D1 and D2 were superior to the others. However, one could not conclude on the suitability of these diets unless and until their responses obtained in the feeding trials are compared with some common market available diets. In this context the present chapter is aimed at comparing the growth performances of *P. sphenops* when fed with D1 and D2 with some commonly used diets.

9.2. Obtaining Ornamental Fish Diets from the Market

Three commonly available aquarium fish diets were obtained from the local market. Two of them had *Spirulina* as the major protein source and one had *Tubifex*. The two diets having *Spirulina* meal had crude protein of about 46 per cent and 32 per cent, hence they were named high protein *Spirulina* (HPS) and low protein *Spirulina* (LPS) respectively. The third one was named *Tubifex*.

9.3. Comparison of Growth Performance

Comparison of growth performance of the selected fishes fed with the market available diets (HPS, LPS and *Tubifex*) and two of the formulated *Oxya* meal added diets (*i.e.*, D1 and D2) were done in terms of average daily growth (per cent ADG) and specific growth rate (per cent SGR). This experiment was conducted for 90 days with a similar set up as discussed in the previous chapter.

9.4. Nutrient Composition of the Market Available Diets

Proximate compositions of the market available diets revealed that dried *Tubifex* cakes contained about 52 per cent protein which is much higher than the optimal need of mollies. On the other hand LPS contained nearly 32 per cent of crude protein which is lower than the need. However, HPS had a crude protein value (46.08 per cent) which is close to the protein requirement of these fishes. Crude fat and energy contents were also high in the *Tubifex* cakes. Carbohydrate and NFE contents were highest in LPS, whereas ash content and P/E ratio was present in highest amount in HPS.

9.5. Growth Performance of *P. sphenops* Fed with Market Available Diets and Selected Oxya Meal Added Diets

It was notable that the results of growth performance between D1 and D2, and the same between the three market available diets (*i.e.* HPS, LPS, *Tubifex*) did not vary significantly (Figures 9.1 and 9.2). However, the results were quite encouraging because both the selected *Oxya* meal added diets showed significantly higher ADG and SGR in comparison with the three selected market available diets.

9.6. Elaboration

From the reports of various scientists it is evident that mollies need a 40 per cent crude protein with a high level of NFE (at least 40 per cent), 5-7 per cent of crude fat and the balance of P/E ratio of 20-22mg/kJ. Commonly available market diets usually have a crude protein level in between 30-50 per cent. Hence while choosing the three most common market available diets three different protein levels were obtained, these were – 46 per cent

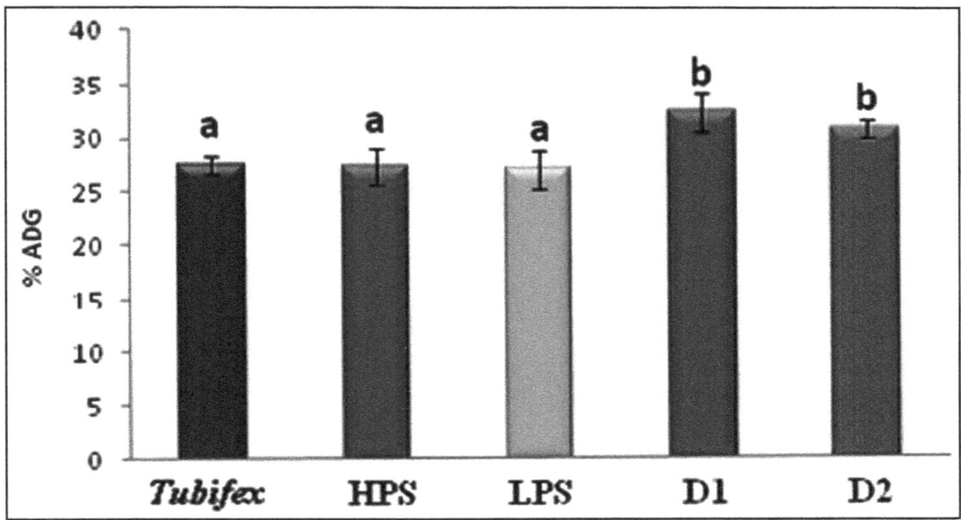

Figure 9.1: Comparison of Average Daily Growth (per cent ADG) of *P. sphenops* when Fed with Market Available Diets and Selected *Oxya* Meal Added Diets (Values with different letter are significantly different P<0.001, ANOVA, Tukey's range test).

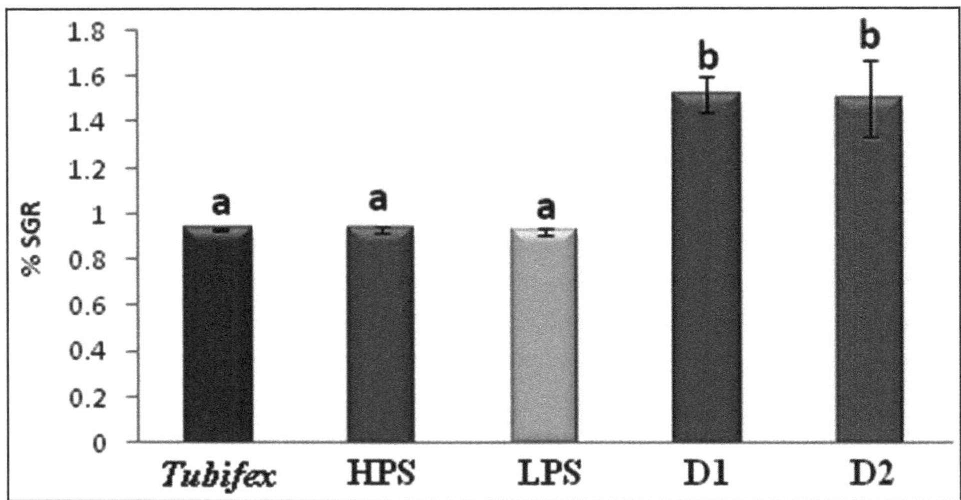

Figure 9.2: Comparison of Specific Growth Rate (per cent SGR) of *P. sphenops* when Fed with Market Available Diets and Selected *Oxya* Meal Added Diets (Values with different letter are significantly different P<0.001, ANOVA, Tukey's range test).

(which is very close to the need of mollies), 32 per cent (lower than the optimum level), and 52 per cent (higher than the optimum level). It is already known that protein plays an important role in fish nutrition. On one hand inadequate protein levels in the diet results growth retardation and loss of weight, on the other hand only a part of an excess of protein supplied in the diet is used for protein synthesis in fishes, while the remaining is transformed into energy. The major adversity regarding excess protein in fish diets is that the breakdown products of metabolism (mainly ammonia) could directly pollute their living environment. Hence, it should be monitored so that the ornamental fishes kept in aquarium can utilise their dietary protein with a great efficiency. In the present work two of the diets were below and above the optimum level of the need of mollies and one was very similar in terms of protein. But for a complete picture one should also focus on the other important factors such as fat, energy, P/E ratio and NFE. It has been already discussed in the previous chapter that NFE is an equally important factor for growth of the live bearer fishes. In the present case two of the three market available diets *i.e. Tubifex* and HPS contained NFE levels which is much lower than the need (*i.e.* 40 per cent and above) of mollies. On the contrary though LPS fulfilled the nutritional need in terms of NFE and P/E ratio, its crude protein level was much lower. This shows that all the three market available diets are somewhat deficient, and this might have lead to a stunted growth of the selected fishes in the feeding trial.

From our discussion up to this level it is clear that the market available diets do not fulfill all the dietary need of mollies, naturally the fishes did not show very good growth when fed on these diets. Here one should remember that these are generalised diets made for many fishes to feed, whereas the selected *Oxya* meal added diets (*i.e.* D1 and D2) are practically balanced diets for mollies. This may be a justification about a higher growth of fishes that was observed when fed on these two diets. In this context it is necessary to discuss whether this *Oxya* meal added diets could be used as a feed for some other fishes also. While surveying literature in this regard we could see that most of the fishes described till date are carnivorous as about 85 per cent belong to this group, while 6 per cent are herbivorous, 4 per cent are omnivorous, 3 per cent are detritivorous and 2 per cent are scavengers or parasites. Protein requirement of all of the carnivorous and omnivorous fishes (*i.e.* about 89 per cent of the total fish group) are higher than the others. These fishes require 40 per cent of protein or above. Investigators who have worked on the dietary need of ornamental fishes reported that a diet with 45 per cent protein and 6 per cent lipid level is needed for the best specific growth rate and feed conversion ratio for cyprinodonts other than mollies, such as swordtails. The omnivorous "tin foil barb" (*Barbodes altus*) and the carnivorous "discus" (*Symphysodon aequifasciata*) also need a high crude protein level of 41.7 per cent and over 45 per cent respectively. All these examples show that the crude protein level of 40 per cent, crude fat level in between 5-6 per cent, NFE level nearly 40 per cent, about 19kJ/g gross energy (GE) and approximately 21mg/kJ of P/E ratio in the formulated *Oxya* meal added diets (Ref. chapter 7) could also be suitable for many other carnivorous and omnivorous fishes in addition to mollies.

9.7. Conclusion

Significantly higher growth was obtained for the selected *Oxya* meal added diets compared to the market available ones (*i.e.* HPS, LPS and *Tubifex*). This proves that the formulated diets are superior to the commonly used diets for black mollies. Moreover it was also noted that these two *Oxya* meal added diets could also be used as generalised feed for some other carnivorous and omnivorous fish groups in addition to mollies.

Chapter 10

Postscript

This book is mainly the journey through which it is revealed that among the two selected short-horn grasshoppers, *Oxya fuscovittata* has the potential to produce a huge biomass due to their high fecundity, fertility and survival rate. It is nutritionally rich that contains a very high crude protein as well as energy. Essential amino acids and important polyunsaturated fatty acids (such as linoleic acid and linolenic acid) are also present in a good amount in this species of insect. Although being a phytophagous insect it contains several anti-nutritional factors like tannin, oxalate and phytin, they are present in a very low titre that is under the tolerance limit of consumption by fishes and human.

Experiment on feeding trial clearly shows that at least 50 per cent of fish meal may be replaced by *Oxya* meal in formulated diets for the ornamental model fish *P. sphenops* (black molly). When the fishes were fed with the formulated diets, over 80 per cent survival was obtained in each of the sets. A clutch of 13 to 16 individuals was obtained for the sets fed with all the formulated diets.

Another feeding trial for estimation of growth of mollies when fed with the two best formulated diets and three more diets commonly available in the ornamental fish market (*i.e.* HPS, LPS and *Tubifex*) further showed encouraging results, as because considerably higher growth was obtained in the fishes that were fed with *Oxya* meal supplemented diets compared to the market available diet fed fishes, which confirmed the formulated diets to be superior. Although we have initially concluded this might be because the diets that were formulated for the present case were specially designed for mollies, whereas the market available diets are rather generalised diets suitable for feeding a range of fish species. At this point it was necessary to assess the efficacy of the *Oxya* supplemented diets to be sure that such diets could also be suitable for other fishes. We have seen from the discussion of chapter 9 that many of the ornamental fish species are carnivorous or omnivorous that could easily accept these *Oxya* meal added diets.

The findings and chapter-wise discussion of the present book finally support the concept of establishing "acridid farms" where *O. fuscovittata* will be mass reared using *Sorghum halepense* as their major food plant. This highly nutritious mass of acridid tissues could be easily and constantly provided to the aqua-feed developers to formulate supplementary diets for ornamental fishes.

Future Works

The present book mainly focused on finding out suitable acridid species to be mass reared in acridid farms with their suitable food plant, and utilisation of that acridid as an alternative protein supplement for ornamental fish. For this purpose the suitability of the formulated diets were evaluated in terms of different indices regarding food consumption, utilisation and growth. However the digestibility test of all the formulated diets has not been carried out. In future it is necessary to evaluate digestibility. In this context feed absorption efficiency (FAE) could be an important index. Digestibility test using chromic oxide as a marker could also be an appropriate measure.

Insects contain chitin. Though many carnivorous and omnivorous fishes have been reported to be able to breakdown chitin, many herbivorous fishes could not do the same. Using chitinase producing bacteria along with acridid meal during feed formulation could make these diets more digestible especially to the herbivorous fishes.

Various authors reported that fermented diets could be more digestible to the fishes and could enhance growth. Hence in future it should be also tried whether acridid added diets could be fermented using yeasts and bacteria and whether those fermented diets could be better than the ones used in this study.

Ideal composition could be found out through mathematical modelling and consequent feeding trial could establish the composition of an ideal diet that could give optimum growth when fed to different fishes.

The present study only focused on the utilisation of acridids as an alternative protein source for "ornamental fish". Further research is necessary to find out whether these insects could be used in the formulated diets of table fishes also.

A detailed study on economic assessment for establishment of an acridid farm is necessary. In this context, economists and biologists both should come to a single platform and evaluate various aspects (including the "cost effective" factor) of acridid farm establishment. Only then these kinds of farms could be made more viable.

Glossary

Acridid: Short-horn grasshoppers under family Acrididae, under order Orthoptera.

AD: Approximate digestibility. It is a measure of how a food is utilised by an individual. It is calculated as the following equation:

$$AD = \frac{F - Fe}{F} \times 100$$

[Where F= average dry weight of food ingested per individual, Fe= average dry weight of feces produced per individual]

ADG: Average daily growth. It is a parameter for measuring growth which shows the average daily trend of growth during the experimental period. It is calculated as the following equation:

$$\%\ ADG = \frac{Wt2 - Wt1}{Wt1 \times T} \times 100$$

[Where Wt1= initial weight, Wt2= final weight, T= number of days].

Anti-nutrient: These are natural or synthetic compounds that hinder the process of absorption of nutrients present in the diet.

Biomass (in dry weight): Total organic matter in dry weight derived from living organisms.

Bivoltine: Insects that can complete two generations in one year.

Chitin: It is an amino sugar which is the major component of insect outer covering (*i.e.* cuticle). It is chiefly made up of N-acetylglucosamine.

Chitinase: Enzyme that can act on chitin as its substrate ultimately breaking it down.

CI: Consumption Index. It is a measure of food consumed by an individual per day. It is calculated as the following equation:

$$CI = \frac{F}{TA}$$

[Where F= average dry weight of food ingested per individual, T= duration of feeding period in days, A= mean dry weight during the feeding period].

Condition factor: It is a measure of an individual fish's health using the ratio of standard length and weight. It is proposed by Fulton in 1904. It assumes that the weight of a fish is proportional to the cube of its standard length. It is estimated by the following equation: $K= 100(W/L^3)$ [Where W is the whole body wet weight in grams and L is the length in centimetres].

Copulation: Insertion or thrusting of penis by the male animal inside its female counterpart for the purpose of reproduction.

Cyanogenic: A substance which is capable of producing hydrogen cyanide.

ECD: Efficiency of conversion of digested food to body substance. It is calculated as the following equation:

$$ECD = \frac{Wt}{F - Fe} \times 100$$

[Where Wt= dry weight gained, F= average dry weight of food ingested per individual, Fe= average dry weight of feces produced per individual]. Also compare with ***ECI***.

ECI: Efficiency of conversion of ingested food to body substance. It is a measure by means of which the utilisation of food materials could be indicated. It is calculated as the following equation:

$$ECI = \frac{Wt}{F} \times 100$$

[Where Wt= dry weight gained, F= average dry weight of food ingested per individual].

Entomophagy: Practice of consumption of insect as food.

FCE: Feed conversion efficiency. It is measure by means of which one can understand how much wet weight has been gained by the individuals with respect to the amount of feed consumed. It is estimated by the following equation: FCE= WWt/F) [Where WWt= wet weight gained, F= Average dry weight of food ingested per individual]. Also see ***FCR***.

FCR: Feed conversion ratio. It is the reciprocal of FCE, *i.e.* 1/FCE.

Fledgling: A young insect that has recently acquired fully developed wings after moulting.

GR: Growth rate. It is calculated as the following equation:

$$\mathbf{GR} = \frac{\mathbf{Wt}}{\mathbf{TA}}$$

[Where Wt= dry weight gained, T= number of days, A= mean dry weight during the feeding period].

Graminivorous: Herbivorous insects that strictly feed on plants under the family Gramineae (presently called Poaceae).

Gonopodium: The modified anal fin used as an intromittent (copulatory) organ, which is chiefly observed in adult male fish of the live bearer fishes (family-Poeciliidae).

Hemimetabolous: Insects that show hemimetaboly. See also **Hemimetaboly.**

Hemimetaboly: Development into an adult through a nymphal stage by means of molting.

Integument: It is the outer covering of insects that comprises epidermis and cuticle.

Minilivestock: It encompasses small indigenous vertebrates and invertebrates which can be produced on a suitable basis for food, animal feed and hence are a source of income.

Multivoltine: Insects that can complete several generations in one year. See also **Univoltine** and **Bivoltine**.

Nymphs: An immature form of insects that are quite similar to the adult form, only they have small size, undeveloped wings and immature gonads

Ornamental fish: Any fish which looks different from our conventional idea of a typical fish. They may be colourful, having various types of stripes or bands, or they could be very different in shape. These fishes may or may not be used as food by various cultures, but are suitable to be kept in aquaria.

Ovipositor: A specialised female organ for laying eggs.

P/E Ratio: It is a ratio which is calculated in terms of amount of protein present in milligrams per kilo Jule of energy in a diet (mg protein KJ^{-1} of energy).

PER: Protein efficiency ratio. It is a measure by means of which one can understand how much protein is utilised for building the body substance. It is estimated by the following equation: PER= WWt/DWp) [Where WWt= wet weight gained, DWp= dry weight of protein in feed].

Primary consumers: Animals that consumes producers (mainly plants), *i.e.* herbivorous animals.

Pronotum: The dorsal plate of prothorax (*i.e.* the segment between head and thorax).

PWG: Percent weight gain. It is measured in terms of percentage of final weight increased from that of the initial weight.

Rearing: Maintenance of live animals in captive condition for generations. This includes feeding and breeding of the animals.

Sclerotisation: Hardening of insect cuticle due to cross-linking of protein chains.

SGR: Specific growth rate. It is a parameter for measuring growth which takes the log value of the initial and final weight. It is calculated as the following equation:

$$\%\mathbf{SGR} = \frac{\log \mathbf{Wt2} - \log \mathbf{Wt1}}{\mathbf{T}} \times 100$$

[Where Wt1= initial weight, Wt2= final weight, T= number of days].

Supplementary diet: A food product that contains a "dietary ingredient" intended to add further nutritional value to (supplement) the diet. A "dietary ingredient" may be one, or any combination, of various nutrients.

Univoltine: Insects that can complete only one generation in one year.

Vitellogenesis: The process of yolk formation via nutrients to be deposited in the egg.

Zoonosis: Any disease or infection that is naturally transmissible from vertebrate animals to human.

References

Aarnink AJA, Keen A, Metz JHM, Speelman L, Verstegen MWA. 1995. Ammonia emission patterns during the growing periods of pigs housed on partially slatted floors. *Journal of Agricultural Engineering Research*. 62: 105–116.

Adeduntan SA. 2005. Nutritional and anti-nutritional characteristics of some insects foraging in Akure forest reserve, Ondo state, Nigeria. *Journal of Food Technology*. 3: 563-567.

Adesulu EA, Mustapha AK. 2000. Use of housefly maggots as a fishmeal replaces in tilapia culture: A recent vogue in Nigeria. In: K. Ftzimmons and J.C. Filho (Eds.). *Proceedings of the Fifth International Symposium on Tilapia Aquaculture*. Rio de Janeiro.

Adeyeye EI. 2011. Fatty acid composition of *Zonocerus variegatus*, *Macrotermes bellicosus* and *Anacardium occidentale* kernel. International Journal of Pharma and Bio Sciences. 2:135-144.

Adriens EL. 1953. Note sur la composition chimique quelques aliments mineurs indigenes du Kwang. *Ann. Soc. Belge Med.Trop*. 33(6):531-534. (In French).

Agbede JO, Aletor VA. 2004. Chemical characterization and protein quality evaluation of leaf protein concentrates from *Glyricidia sepium* and *Leucaena leucocephala*. *International Journal of Food Science and Technology*. 39: 253-361.

Ajani EK, Nwanna, LC, Musa BO. 2004. Replacement of fishmeal with maggot meal in the diets of Nile tilapia, *Oreochromis niloticus*. *World Aquaculture*. 35: 52–54.

Akand AM, Hassan MR, Habib MAB. 1991. Utilisation of carbohydrate and lipid as dietary energy sources by stinging catfish *H. fossilis* (Bloch). In: S. DeSilva (Ed.). Fish nutrition research in Asia. Proceedings of the fourth Asian fish nutrition workshop. Asian Fish Society, special publication, Manila, Philippines. Asian Fisheries Society.

Akegbejo-Samson Y. 1999. Growth response and nutrient digestibility by *Clarius gariepinus* fed on various level of dietary periwinkle flesh as replacement for fish meal in low cost diet. *Applied Tropical Agriculture*. 4: 37-43.

Akinnawo O, Ketiku AO. 2000. Chemical composition and fatty acid profile of edible larva of *Cirina forda* (Westwood). *African Journal of Biomedical Research*. 3: 93-96.

Akinwande AA, Ugwumba AAA, Ugwumba OA. 2002. Effects of replacement of fish meal with maggot meal in the diet of *Clarias gariepinus* (Burchell 1822) fingerlings. *The Zoologist*. 1: 41-46.

Alegbeleye WO, Oresegun A. 1998. Nutritive value of three terrestrial lumbrid worms for *Oreochromis niloticus*. Sustainable Utilisation of Aquatic/Wetland Resources. Selected Papers from 9th /10th Annual Conference of the Nigerian Association for Aquatic Sciences.

Alemla AM, Singh HK. 2004. Utilisation of insect as human food in Nagaland. *Indian Journal of Entomology*. 66: 308–310.

Ali MZ, Jauncey K. 2005. Approaches to optimizing dietary protein to energy ratio for African catsh *Clarias gariepinus* (Burchell, 1822). *Aquaculture Nutrition*, 11: 95–101.

Ali SA. 2008. Beneficial feed additives from bioresources for aquafeeds. Proceedings of 8th Indian Fisheries Forum, Kolkata.

Anand H. 2009. Evaluation of the feed value of acridids and its utilisation in formulating economic poultry feed. Unpublished thesis submitted to Visva-Bharati University, India.

Anand H, Das S, Ganguly A, Haldar P. 2008a. Biomass production of acridids as possible animal feed supplement. *Journal of Environment and Sociobiology* 5: 181-190.

Anand H, Ganguly A, Haldar P. 2008b. Potential value of Acridids as high protein supplement for poultry feed. *International Journal of Poultry Science* 7: 722-725.

Ananthakrishnan TN, Dheelipan K, Padmanabhan B. 1985. Behavioral responses in terms of feeding and reproduction in some grasshoppers (Orthoptera: Insecta). *Proceedings of Indian Academy of Science. (Animal science)*. 94:443-461.

Anthonio HO, Isoun M. 1982. Nigerian cookbook.1st ed. Macmillan Press, London.

AOAC (Association of official analytical chemists). 2000. Official methods of analysis. 17th edition. Association of Official Analytical Chemists, Arlington, USA.

Arockiaraj AJ, Muruganandam M, Marimuthu K, Haniffa MA. 1999. Utilisation of carbohydrates as a dietary energy source by stripped murrel *Channa striatus* (Bloch) fingerlings. *Acta Zoologica Taiwanica*. 10:103-111.

Ashiru MO.1988. The food value of the larvae of *Anaphe venata* Butler, Lepidoptera: Notodoutidae. *Ecology of Food and Nutrition*. 22:313-322.

Awmack CS, Leather SR. 2002. Host plant quality and fecundity in herbivorous insects. *Annual Review of Entomology*. 47: 817-844.

Bachstez A, Aragon A. 1945. Notes on Mexican drugs, plants and foods. *Journal of American Pharmaceutical Association*. 34: 170-172.

Bailey M, Sandford G. 2001. The ultimate aquarium. 2nd edition. Anness Publishing Ltd. London.

Banerjee GC. 1992. Poultry. 3rd edition. Oxford and IBH Publishing Co. Pvt. Ltd, New Delhi.

Banjo AD, Lawa OA, Songonuga EA. 2006. The nutritional value of fourteen species of edible insects in southwestern Nigeria. *African Journal of Biotechnology*. 5:298-301.

Beets WC. 1997. The need for an increased use of small and mini-livestock in integrated smallholder farming systems. *Ecology of Food and Nutrition*. 36:237–245.

Berenbaum MR. 1993. Sequestered plant toxins and insect palatability. *Food Insects Newsletter*. 6:1, 6, 7, and 9.

Bernays EA, Bright KL. 1991. Dietary mixing in grasshoppers: switching induced by nutritional imbalances in foods. *Entomologia Experimentalis et Applicata* 61: 247-253.

Bettelheim FA, Landesberg JM. 1997. Laboratory experiments for general, organic and biochemistry. Saunders College Publishing, USA.

Bodenheimer FS. 1951. Insects as Human Food. W. Junk, The Hague.

Bondari K, Sheppard DC. 1981. Soldier fly larvae as feed in commercial fish production. *Aquaculture*. 24: 103-109.

Boyd CE. 1990. Water quality in ponds for aquaculture. Alabama Agricultural Experimentation Station, Auburn University, Alabama.

Chen XM, Feng Y. 1999. Edible insects in China. China Science and Technology Press, Beijing.

Chen XM, Ying F, Zhiyong C. 2009. Common edible insects and their utilisation in China. *Entomological Research*. 39:299-303.

Cherikoff V, Isaacs J. 1989. The bushfood handbook. Ti Tree Press, Sydney.

Chinn M. 1945. Notes pour l'etude de l'alimentation des indigenes de la province de Coquitamatville. *Ann. Soc. Belge Med.Trop*. 25: 57-149 (In French).

Chong ASC, Hashim R, Ali AB. 2000. Dietary protein requirements for discus (*Symphysodon* spp.). *Aquaculture Nutrition*. 6: 275–278.

Cole NA, Clark RN, Todd RW, Richardson CR, Gueye A, *et al.* (2005). Influence of dietary crude protein concentration and source on potential ammonia emissions from beef cattle manure. *Journal of Animal Science*. 83: 722–731.

Comby B. 1990. Delicious Insects. Editions Jouvence, Geneva.

Cowey CB, Sargent JR. 1979. Nutrition. In: W. S. Hoar and J. Randall (Eds.), Fish physiology. NY: Academic Press, New York.

Craig S, Helfrich LA. 2009. Understanding fish nutrition, Feeds and feeding. Verginia cooperative extension. Publication No. 420-256.

Das A, Das S, Haldar P. 2002. Effect of food plants on growth rate and survivality of *Hieroglyphus banian* (Fabricius) (Orthoptera: Acridoidea), a major paddy pest in India. *Applied Entomology and Zoology*. 37: 207-212.

DeFoliart GR. 1975. Insects as source of protein. Bulletin of the Entomological Society of America. 21: 161-163.

DeFoliart GR. 1989. The human use of insects as food and as animal feed. *Bulletin of the Entomological Society of America*. 35:22-35.

DeFoliart GR. 1991. Insect fatty acids: similar to those of poultry and fish in their degree of unsaturation, but higher in the polyunsaturates. *Food Insects Newsletter*. 4: 1-4.

DeFoliart GR. 1992. Insects as food: Nutritional and economical aspects. *Crop Production*. 11: 395-399.

DeFoliart GR. 1999. Insects as food: why the Western attitude is important. *Annual Review of Entomology*. 44: 21–50.

DeFoliart GR, Finke MD, Sunde ML. 1982. Potential value of the Mormon cricket (Orthoptera: Tettigoniidae) harvested as a high-protein feed for poultry. *Journal of Economic Entomology*. 75: 848-852.

DeFoliart GR, Nakagaki B, Sunde ML. 1987. Protein quality of the house cricket *Acheta domesticus* when fed to broiler chicks. *Poultry Science*. 66: 1367-1371.

DeGuevara LO, Padilla P, Garcia L, Pino JM, Ramos-Elorduy J. 1995. Amino acid determination in some edible Mexican insects. *Amino Acids*. 9: 161-173.

Delisle J, Hardy M. 1997. Male larval nutrition influences the reproductive success of both sexes of the spruce budworm, *Choristoneura fumiferana* (Lepidoptera: Tortricidae). *Functional Ecology*. 11: 451–463.

Denulat E. 1986. The effects of various diets on chitinase and b-glucosidase activities and the condition of cod, *Gadus morhua* (L.). *Journal of Fish Biology*. 28: 191–197.

DeSchriver MM, Riddick CC. 1990. Effects of Watching Aquariums on Elders' Stress. *Anthrozoos*. 4: 44-48.

De Silva SS, Anderson A. 1998. Fish nutrition in aquaculture. 1st edition. Chapman and Hall, London.

Despins JL. 1994. Feeding behavior and growth of turkey poults fed larvae of the dakling beetle (*Alphitobius diaperinus*). *Poultry Science*. 74: 1526-1533.

Dey A, Hazra AK. 2003. Diversity and distribution of grasshopper fauna of greater Kolkata with notes on their ecology. In: Memoires of the Zoological Survey of India, Vol. 19. Issue 3.

Diss AL, Kumkel JG, Montgomery ME, Leonard DE. 1996. Effects of maternal nutrition and egg provisioning on parameters of larval hatch, survival and dispersal in the gypsy moth *Lymantria dispar* L. *Oecologia*. 106: 470–477.

Dreyer JJ, Weameyer AS. 1982. On the nutritive value of mopanie worms. *South African Journal of Science*. 78: 33-35.

Dufour DL. 1987. Insects as food: a case study from the Northwest Amazon. *American Anthropologist*. 89: 383-397.

Earle KE. 1995. The nutritional requirements of ornamental fish. The *Veterinary Quarterly*. 17 (Supplement 1): S53-S55.

Edwards NE, Beck AM. 2002. Animal-Assisted Therapy and Nutrition in Alzheimer's Disease. *Western Journal of Nursing Research*. 24: 697-712.

Ekop AS. 2004. Effects of Processing on the Chemical Composition of Maize. 24th Annual Conference of Nigerian Society of Biochemistry and Molecular Biology. University of Calabar.

Ekop EA, Udoh AI, Akpan PE. 2010. Proximate and anti-nutrient composition of four edible insects in Akwa Ibom state, Nigeria. *World Journal of Applied Science and Technology*. 2: 224 – 231.

Elangovan A, Shim KF. 2000. The influence of replacing fish meal partially in the diet with soybean meal on growth and body composition of juvenile tin foil barb (*Barbodes altus*). *Aquaculture*. 189: 133–144.

Ewen AB, Hinks CF. 1986. Rearing a non-diapause strain of the migratory grasshopper *Melanoplus sanguinipes* (F) (Orthoptera: Acrididae) in the laboratory. *Proceedings of Pan American Acridological Society* 4: 169-173.

Eyo JE, Ezechie CU. 2004. The effects of rubber (*Havea brasiliensis*) seed meal based diets on diets acceptability and growth performance of

Heterobranchus bidorsalis (male) *Clarias gariepinus* (female) hybrid. *Journal of sustainable Tropical Agricultural Research*. 10: 20-25.

Fabbri C, Valli L, Guarino M, Costa A, Mazzotta V. 2007. Ammonia, methane, nitrous oxide and particulate matter emissions from two different buildings for laying hens. *Biosystems Engineering*. 97: 441–455.

Fagonee Y. 1983. Inclusion of silkworm pupae in poultry rations. *Journal of Tropical Veterinary*. 1: 91-96.

Falayi BA, Balogun AM, Adebayo OT, Madu CT, Eyo AA. 2004. Comparison of seven locally prepared starches with sodium carboxyl methyl-cellulose for water stability in African catfish (*Clarias gariepinus*) feeds. *Journal of Sustainable Tropical Agricultural Research*. 9: 104-108.

Fanny P, Ravikumar T, Muralirangan MC, Sanjayan KP. 1999. Influence of host plants on the duration of post-embryonic development and food utilisation of *Oxya nitidula* (Walker) (Orthoptera: Acrididae). *Journal of Orthoptera Research*. 8:119-124.

Farnando AA, Phang VPE, Chan SY.1991. Diets and feeding regimes of pociliid fishes in Singapore. *Asian Fisheries Science*. 4:99-107.

Fasakin EA, Balogun AM, Ajayi OO. 2003. Evaluation of full-fat and defatted maggot meals in the feeding of Clariid catfish *Clarias gariepinus* fingerlings. *Aquaculture Research*. 34: 733–738.

Fasoranti JO, Ajiboye DO. 1993. Some edible insects of Kwara State, Nigeria. *American Entomologist*. 39: 113-116.

Feng Y, Chen XM, Ye SD, Wang SY, Chen Y, Wang ZL. 1999a. Note on two edible insects species in Yunnan and their nutritious analysis. In: XM. Chen (Ed.) Research and Development of Resource Insects. Yunnan Science and Technology Press, Kunming.

Feng Y, Chen XM, Ye SD, Wang SY, Chen Y, Wang ZL. 1999b. Records of four species edible insects in Homoptera and its nutritive elements analysis. *Forest Research*. 12: 515–518.

Feng Y, Chen XM, Wang SY, Ye SD, Chen Y. 2000a. The nutritive elements analysis of bamboo insect and review on its development and utilisation value. *Forest Research*. 13: 188–191.

Feng Y, Chen XM, Wang SY, Ye SD, Chen Y. 2000b. The common edible insects of Hemiptera and its nutritive value. *Forest Research*. 13: 612–620.

Feng Y, Chen XM, Wang SY, Ye SD, Wang ZL. 2001a. Studies on the nutritive value and food safety of *Ericerus pela* eggs. *Forest Research*. 14: 322–327.

Feng Y, Chen XM, Wang SY, Ye SD, Chen Y. 2001b. Three edible Odonata species and their nutritive value. *Forest Research*. 14: 421–424.

Feng Y, Chen XM, Ye SD, Wang SY, Chen Y, Wang ZL. 2001c. The common edible species of wasps in Yunnan and their value as food. *Forest Research*. 14: 578–581.

Finke MD, Sunde ML, DeFoliart GR. 1985. An evaluation of the protein quality of Mormon Crickets when used as high protein feedstuff for poultry. *Poultry Science*. 64: 708-712.

Fischl J. 1960. Quantitative colorimetric determination of tryptophan. *Journal of Biological Chemistry*. 235: 999-1001.

Fox CW, Nilsson JA, Mousseau TA. 1997. The ecology of diet expansion in a seed-feeding beetle: pre-existing variation, rapid adaptation and maternal effects? *Evolutionary Ecology*. 11: 183–194.

Francis-Floyd R. 2011. Dissolved Oxygen for Fish Production. Fisheries and Aquatic sciences Department, Florida Cooperative Extension Service, Institute of Food and Agricultural Sciences, University of Florida. Publication No. FA 27.

Froese R, Pauly D. 2011. Fish Base. World Wide Web electronic publication. www.fishbase.org, version.

Furuichi M, Yone Y. 1981. Availability of carbohydrate in nutrition of carp and red sea breem. *Bulletin of the Japanese Society of Sceintific Fisheries*. 48: 945-948.

Ganguly A, Chakravorty R, Das M, Gupta M, Mandal DK, Haldar P, Ramos-Elorduy J, Moreno JMP. 2013. A preliminary study on the estimation of nutrients and anti-nutrients in *Oedaleus abruptus* (Thunberg) (Orthoptera:Acrididae). *International Journal of Nutrition and Metabolism*. 5: 50-56.

Ganguly A, Chakravorty R, Haldar P. 2013. Assessment of consumption, utilisation and growth of *Oedaleus abruptus* (Thunberg) and *Spathosternum prasiniferum prasiniferum* (Walker) (Orthoptera: Acrididae) fed with various food plants in laboratory conditions. *Annales de la Societe Entomologique de France*. 49: 160-171.

Ganguly A, Chakravorty R, Sarkar A, Haldar P. 2010. Johnsongrass [*Sorghum halepense* (L.) Pers.]: A potential food plant for attaining higher grasshopper biomass in acridid farms. *Philippine Agricultural Scientist*, 93: 329-336.

Ganguly A, Chakravorty R, Sarkar A, Mandal DK, Haldar P, et al. 2014. A preliminary study on *Oxya fuscovittata* (Marschall) as an alternative nutrient supplement in the diets of *Poecillia sphenops* (Valenciennes). PLoS ONE 9(11): e111848. doi:10.1371/journal.pone.0111848.

Gatlin DM. 2010. Principles of fish nutrition. Southern Regional Aquaculture Center. Publication No. 5003.

Ghaly AE. 2009. The Black Cutworm as a potential human food. American *Journal of Biochemistry and Biotechnology* 5: 210-220.

Ghosh AK, Naskar AK, Jana ML, Khowala S, Sengupta S. 1995. Purification and characterization of an amyloglucosidase from *Termitomyces clypeatus* that liberates glucose from xylan. *Biotechnology Progress*. 11: 452-456.

Gope B, Prasad B. 1983. Preliminary observations on the nutritional value of some edible insects of Manipur. *Journal of Advance Zoology*. 4:55–61.

Gorham JR. 1979. The signicance for human health of insects in food. *Annual Review of Entomology*. 24: 209–224.

Guillaume J. 1997. Protein and amino acids. In: L.R. D'Abramo, D.E. Conclin, D. M. Akiyama (Eds.). Crustacean nutrition, advances in world aquaculture, vol. 6. World Aquaculture Society, Louisiana, USA.

Gullan PJ, Cranston PS. 2005. The insects: an outline of entomology: Blackwell Publishing.

Gunn DL. 1960. The biological background of Locust control. *Annual Review of Entomology*. 5: 279-300.

Gupta PC, Khatta VK, Mandal AB. 1988. Analytical techniques in animal nutrition. Haryana Agricultural University, Hisar.

Haldar P, Bhandari KP, Nath S. 1995. Observations on food preferences of an Indian grasshopper *Acrida exaltata* (Walker) (Orthoptera:Acrididae:Acridinae). *Journal of Orthoptera Research*. 4: 57-59.

Haldar P, Das A, Gupta RK. 1999. A laboratory based study on farming of an Indian grasshopper *Oxya fuscovittata* (Marschall) (Orthoptera: Acrididae). *Journal of Orthoptera Research*. 8: 93-97.

Haldar P, Nandi NC. 1997. Feasibility of grasshopper farming in W.B. - A review on laboratory based studies. *Journal of Interacademica*. 1: 160-161.

Hasan MR, Moniruzzaman M, Farooque AMO. 1990. Evaluation of leucaena and water hyacinth leaf meal as dietary protein sources for the fry of Indian major carp, *Labeo rohita* (Hamilton). In: R. Hirano and I. Hanyu (Eds). *Second Asian Fisheries Forum*. Asian Fisheries Society. Manila, Philippines.

Hassan LG, Umar KJ, Dangoggo SM, Maigandi AS. 2011. Anti-nutrient composition and bioavailability prediction as exemplified by calcium, iron and zinc in *Melocia corchorifolia* leaves. *Pakistan Journal of Nutrition*. 10: 23-28.

Hazra AK, Dey A, Hazra N.1997. Studies on ecology and taxonomy of grasshopper (Acridoidea) fauna along the Damoder river bed agro-ecosystem of West Bengal. In: M.P. Sinha (Ed.). *Recent Advances in Ecobiological Research*. Vol 1. Sundeep Prakashan, New Delhi.

Heidinger RC. 1971. Use of ultraviolet light to increase the availability of aerial insects to caged bluegill sunfish. *Progressive Fish-Culturist*. 33:187-192.

He JZ, Tong Q, Huang XH, Zhou ZH. 1999. Nutritive composition analysis of moths of *Dendrolimus houi* Lajongquiere. *Entomological Knowledge* 36: 83–86.

Hewitt GB. 1969. Twenty six varieties of forage crops evaluated for resistance to feeding by *Melanoplus sanguinipes* (F). *Annals of the Entomological Society of America*. 62: 737-741.

Hinks CF, Erlandson MA. 1994. Rearing grasshoppers and locusts: review, rationale and update. *Journal of Orthoptera Research*. 3: 1-10.

Hinks CF, Olfert O. 1992. Cultivar resistance to grasshoppers in temperate cereal crops and grasses: a review. *Journal of Orthoptera Research*. 1: 1-9.

Holt VM. 1969. Why not Eat Insects? International Specialized Book Service Inc., London.

Hossain MH, Ahammad MU, Howlider MAR. 2003. Replacement of fish meal by broiler offal in broiler diet. *International Journal of Poultry Science*. 2:159-163.

Hu C. 1996. Resource Insects and Utility. China Agriculture Press, Beijing.

Ifie I, Emeruwa CH. 2011. Nutritional and anti-nutritional characteristics of the larva of *Oryctes monoceros*. *Agriculture and Biology Journal of North America*. 2: 42-46.

Jamil K, Abbas G, Akhtar R, Hong L, Zhenxing L. 2007. Effects of replacing fish meal with animal bi-products meal supplementation in diets on the growth and nutrient utilisation of mangrove red snapper. *Journal of Ocean University of China* 6: 292-298.

Jeuniaux C, Dandrifosse G, Micha JC. 1982. Caracte' res et e´ volution des enzymes chitinolytiques chez les verte´ bre´ s infe´ rieurs. *Biochemical Systematics and Ecology*. 10: 365–372. (In French).

Joren A. 2001. Nutritional needs and control of feeding. In: G.L. Cunningham, M.W. Sampson, (Eds.). Grasshopper integrated pest management user handbook, Washington DC.

Joren A, Gaines SB. 1990. Population dynamics and regulation in grasshoppers. In: R.E. Chapman, A. Joren (Eds). Biology of Grasshoppers. John Wiley and Sons, New York.

Joshi PS, Rao PV, Mitra A, Rao BS. 1980. Evaluation of de-oiled silkworm pupae-meal on layer performance of broiler chicks. *Indian Journal of Animal Science*. 50: 979-982.

Kalita P, Mukhopadhyay PK, Mukherjee AK. 2007. Evaluation of the nutritional quality of four unexplored aquatic weeds from North East India for the formulation of cost-effective fish feeds. *Food Chemistry*. 103: 204-209.

Kim K, Kayes TB, Amundson CH. 1991. Purified diet development and re-evaluation of the dietary protein requirement of fingerling rainbow trout (*Oncorhyncus mykiss*). *Aquaculture*. 96: 57-67.

Kirby WF. 1914. The fauna of British India including Ceylon and Burma. Orthoptera, Vol. I (Acrididae). Taylor and Francis, Red Lion Court, Fleet st. London.

Kodondi KK, Leclerc M, Bourgeasy-Causse P, Gaudin H. 1987a. Interet nutritionnel de chenilles d'attacides du Zaire. Composition et valeur nutritionnalle. *Cahiers de Nutrition et de Dietetique*. 22: 473-477. (In French).

Kodondi KK, Leclerc M, Gaudin HF. 1987b. Vitamin estimations of three edible species of Attacidae caterpillars from Zaire. *International Journal of Vitamin and Nutrition Research*. 57: 333-334.

Krogdahl A, Hemre GI, Mommsen TP. 2005. Chitinase adopted from Carbohydrates in fish nutrition: digestion and absorption in postlarval stages. *Aquaculture Nutrition*. 11: 103–122.

Kruger DP, Britz PJ, Sales J. 2001a. The influence of live feed supplementation on growth and reproductive performance of swordtail (*Xiphophorus helleri* Heckel 1848). *Aquarium Sciences and Conservation*. 3:275-280

Kruger DP, Britz PJ, Sales J. 2001b. Influence of varying dietary protein content at three lipid concentrations on growth characteristics of juvenile swordtails (*Xiphophorus helleri* Heckel 1848). *Aquarium Sciences and Conservation*. 3: 275–280.

Lacey RE, Redwine JS, Parnell CB. 2002. Emission factors for broiler production operations: A stochastic modeling approach. ASAE Annual International Meeting/CIGR 15th World Congress Chicago, Illinois.

Lagler, K.F. 1956. Freshwater Fishery Biology. W.C.Brown Company, Dubuque, Lowa.

Landry SV, DeFoliart GR, Sunde ML. 1986. Larval protein quality of six species of Lepidoptera (Saturniidae, Sphingidae, Noctuidae). *Journal of Economic Entomology*. 79: 600-604.

Leary DF, Lovell RT. 1975. Value of fiber in production type diets for channel catfish. *Transactions of the American Fisheries Society*. 104: 328-332.

Ledger J. 1987. The eighth plague returneth! The locusts are coming. *African Wildlife*. 41: 201-210.

Lim LC, Sho A, Dhert P, Sorgeloos P. 2001. Production and application of on-growth *Artemia* in fresh water ornamental fish farm. *Aquaculture Economics and Management*. 5: 211-228.

Lindsay GJH, Walton MJ, Adron JW, Fletcher TC, Cho CY, Cowey C.B. 1984. The growth of rainbow trout (*Salmo gairdneri*) given diets containing chitin and its relationship to chitinolytic enzymes and chitin digestibility. *Aquaculture*. 37: 315–334.

Ling KH, Lim LY.2006. The status of ornamental fish industry in Singapore. *Singapore Journal of Primary Industries*. 32: 59-69

Lockwood JA. 1993. Environmental issues involved in biological control of rangeland grasshoppers (Orthoptera: Acrididae) with exotic agents. *Environmental Entomology*. 22: 503-518.

Lokeshwari RK, Shantibala T. 2010. A Review on the Fascinating World of Insect Resources: Reason for thoughts. *Psyche*. 2010:1-11.

Lomer CJ, Bateman P, Johnson DL, Langewald J, Thomas M. 2001. Biological control of locusts and grasshoppers. *Annual Review of Entomology*. 46: 667-702.

Looy H, Wood JR. 2006. Attitudes toward invertebrates: Are educational "bug banquets" effective? *The Journal of Environmental Education*. 37: 37–48.

Love RM.1980. The chemical biology of fishes. Academic press, London.

Lovell T. 1998. Nutrition and feeding of fish. Springer.

Lu Y, Wang DR, Han DB, Zhang ZS, Zhang CH. 1992. Analysis of the patterns and contents of amino acids and fatty acids from *M. annandalei* (Silvestri) and *M. barneyi* Light. *Acta Nutrimenta Sinica*. 14: 103–106.

Macartney A. 1996. Ornamental fish nutrition and feeding. In: N.C. Kelly, J.M. Wills (Eds.), Manual of companion animal nutrition and feeding. British Small Animal Veterinary Association, Gloucestershire, UK.

Macfarlane WV. 1978. Aboriginal desert hunter/gatherers in transition. In: B.S. Hetzel, H.J. Smith (Eds.), The nutrition of aboriginals in the relation to ecosystems of central Australia. Canberra, Melbourne.

Macfarlane JH, Thorsteinson AJ. 1980. Development and survival of the two striped grasshopper *Melanoplus bivittatus* (Say) (Orthoptera: Acrididae) on various single and multiple plant diets. *Acrida*. 9: 63-76.

Malaisse F, Parent G. 1980. Les chenilles comentibles du Shaba Meridional (Zaire). *Naturalistes Belges*. 63: 2-24. (In French).

Malaisse F, Parent G. 1991. Dissponibilite des produits sauveges comestibles de la regione zambezienne: Ecologie et valeur alimentaire des insects et des reptiles. Resumen des communications L'alimentation en foret tropicale: Interactions bioculturelles et applications au development. *Symposium International UNESCO*. (In French).

McCaffery AR. 1975. Food quality and quantity in relation to egg production in *Locusta migratoria migratoriodes*. Journal of Insect Physiology. 21:1551-1558.

McCaffery AR, Cook G, Page WW, Perfect TJ. 1978. Utilisation of food by *Zonocerus variegatus* (L.) (Orthoptera; Pyrgomorphidae). *Bulletin of Entomological Research*. 68: 589-606.

McHargue JS. 1917. A study of the proteins of certain insects with refence to their value as a food for poultry. *Journal of Agricultural Research*. 10: 633-637.

Mehrotra KN, Rao PJ, Farooqi TNA. 1972. The consumption, digestion and utilisation of food by locusts. *Entomologia Experimentalis et Applicata*. 15: 90-96.

Melo V, Garcia M, Sandoval H, Jiménez HD, Calvo C. 2011. Quality proteins from edible indigenous insect food of Latin America and Asia. *Emirates Journal of Food and Agriculture*. 23: 283-289.

Merkowsky AJ, Handcock AJ, Newton SH. 1977. Attraction of aerial insects as a fish food supplement. *Arkansas Academy of Science Proceedings*. 31: 75-76.

Mertinez MH, Ramos-Elorduy J, Pino-Moreno JM, Acosta-Castaneda C. 2008. Evaluation of diets including nymphs of *Periplaneta Americana*. *Ciencia Pesquera*. 16:25-30. (In Spanish with English abstract).

Mignon J. 2002. L'entomophagie: une question de culture? Tropicultura. 20: 151–156. (In French)

Mitsuhashi J. 1992. Edible Insects of the World. Kokinshoin, Tokyo.

Miura K, Ohsaki N. 2004. Diet mixing and its effect on polyphagous grasshopper nymphs. *Ecological Research*. 19: 269-274.

Mohanty SM, Rath SC. 2008. Terrestrial and plant ingredient sources accessibility for aquafeeds. *Proceedings of 8th Indian Fisheries Forum*, Kolkata.

Moyle PB, Chech JJ. 1996. Fishes— An introduction to ichthyology. 3rd edn. Prentice Hall Inc., Englewood Cliffs, NJ.

Munro HN, Crim MC. 1988. The proteins and amino acids. In: M.E. Shils, V.R. Young (Eds.), Modern nutrition in health and disease. Lea and Febiger. Philadelphia, Pensylvennia.

Muthukrishnan J, Delvi MR. 1974. Effect of ration levels on food utilisation in the grasshopper, *Poecilocerus pictus*. *Oecologia*. 16: 227-236.

Nakagaki BJ, DeFoliart GR. 1991. Comparison of diets for mass-rearing *Acheta domesticus* (Orthoptera: Gryllidae) as a novelty food, and comparison of food conversion efficiency with values reported for livestock. *Journal of Economic Entomology*. 84: 891–896.

Nakagaki BJ, Sunde ML, DeFoliart GR. 1987. Protein quality of the house cricket, *Acheta domesticus*, when fed to broiler chicks. *Poultry Science*. 66: 1367-1371.

National Research Council (NRC). 1983. Nutrient requirements of fish. National Academy press, Washington.

Naylor RL, Goldburg RJ, Primavera JH, Kautsky N, Beveridge MCM, Clay J, Folke C, Lubchenco J, Mooney H, Troell M. 2000. Effect of aquaculture on world fish supplies. *Nature*. 405:1017-1024.

Nnaji CJ, Okoye FC. 2005. Substituting fish meal with grasshopper meal in the diet of *Clarias gariepinus* fingerlings. *19th Annual Conference of the Fisheries Society of Nigeria* (FISON).

Nyirenda J, Mwabumba M, Kaunda E, Sales J. 2000. Effect of substituting animal protein source with soybean meal in diets of *Oreochromis karongae* (Trewavas). *The ICLARM Quarterly*. 23: 13-15.

Nzekwu AN, Akingbohungbe AE. 2002. The effects of various host plants on nymphal development and egg production in *Oedaleus nigeriensis* Uvarov (Orthoptera: Acrididae). *Journal of Orthoptera Research*. 11: 185-188.

Ofojekwu PC, Kigbu AA.2002. Effect of substituting fish meal with sesame *Sesamum indicum* (L.) cake on growth and food utilisation of the Nile tilapia *Oreochromis niloticus*. *Journal of Aquatic Sciences*. 17. 45-49.

Ogunji JO, Kloas W, Wirth M, Schulz C, Rennert B. 2006. Housefly maggot meal (Magmeal): An emerging substitute of fishmeal in tilapia diets. Conference on international agricultural research for development; Deutscher Tropentag, Bonn Germany http://www.tropentag.de/2006/abstracts/full/76.pdf.

Ogunji JO, Toor SR, Schulz C, Kloas W. 2008. Growth performance, nutrient utilisation of Nile tilapia *Oreochromis niloticus* fed housefly maggot meal (Magmeal) Diets. *Turkish Journal of Fisheries and Aquatic Sciences.* 8: 141-147.

Ogunlade I, Ilugbiyin A, Osasona AI. 2011. A comparative study of proximate composition, antinutrient composition and functional properties of *Pachira glabra* and *Afzelia africana* seed flours. *African Journal of Food Science.* 5: 32 - 35.

Ojewola GS, Udom SF. 2005. Chemical evaluation of the nutrient composition of some unconventional animal protein sources. *International Journal of Poultry Science.* 4:745-747.

Olfert O, Hinks CF, Westcott ND, Crowle WL, Dziadyk DA, Duczek L. J. 1988. Differential feeding by grasshoppers and levels of foliar diseases in various cultivars of spring cereals. *Crop Protection.* 7: 338-343.

Omotoso OT. 2006. Nutritional quality, functional properties and anti-nutrient composition of the larva of *Cirina forda* (Westwood) (Lepidoptera: Saturniidae). *Journal of Zhejiang University Science* B. 7: 51-55.

Omoyinmi GAK, Fagade SO, Adebisi AA. 2005. Nutritive value invertebrates cultured under laboratory conditions. *The Zoologist* 3: 33-39.

Oonincx DG, van Itterbeeck J, Heetkamp MJ, van den Brand H, van Loon JJ, van Huis A. 2010. An exploration on greenhouse gas and ammonia production by insect species suitable for animal or human consumption. *PLoS ONE.* 5: e14445.

Osuigwe DI, Obiekezie AI, Onuoha GC. 2006. Effects if jackbean seed meal on the intestinal mucosa of juvenile *Heterobranchus longifilis*. *African Journal of Biotechnology.* 13: 1294-1298.

Pannevis MC. 1993. Nutrition of ornamental fish. In: I.H. Burger (Ed.), The Waltham Book of Companion Animal Nutrition. Pergamon Press, Oxford.

Pannevis MC. Earle KE. 1995. Nutrition of ornamental fish. Wien. Tierärztl. Monatsschr. 82: 96–99. (In German).

Paoletti MG. 2005. Ecological implications of minilivestock: potential of insects, rodents, frogs, and snails. Oxford IBH Publishing, Eneld, New Hampshire, USA.

Parker R. 2011. Aquaculture Science. Cengage Learning.

Pérez RM, Ramos-Elorduy J, Yescas G, Munoz JL. 1989. Aislamiento de fenoxazina a partir del insecto *Pachilis gigas* (Insecta- Hemiptera-Coreidae). *Acta Mexicana de Ciencia y Tecnologia* 4: 21-24. (In Spanish).

Phelps RJ, Struthers JK, Moyo J.L. 1975. Investigations into the nutritive value of *Macrotermes falciger* (Isoptera-Termitidae). *Zoologica Africa*. 10: 123–132.

Prain D. 1963. Bengal plants. Botanical survey of India, Calcutta.

Prasad B. 2007. Insect diversity and endemic bioresources. In Endemic Bio-Resources of India- Conservation and Sustainable Development with Special Reference to North East India. Dehra Dun, India.

Price PW, Bouton CE, Gross P, Mcpherson BA, Thompson JN, Weis AE. 1980. Interactions among three trophic levels: Influence of plants on interactions between insect, herbivores and natural enemies. *Annual Reviews of Entomology and Systematics* 11: 11-65.

Quin PJ. 1959. Foods and Feeding Habits of the Pedi. Witwatersrand University, Johannesburg, Republic of South Africa.

Rahman AKM. 2001. Food consumption and utilisation of the grasshopper *Chrotogonus lugubris* Blanchard (Orthoptera: Acridoidea, Pyrgomorphidae) and its effect on the egg deposition. *Journal of Central European Agriculture*. 2: 263-270.

Raj VSD, Jesily S. 1996. The interrelationship of NH_3–N excretion with changing pH, selected metallic ions and food in *Puntius tetrazona*. *Journal of Ecobiology*. 8: 9–15.

Ramos-Elorduy J. 1998. Creepy crawly cuisine: the gourmet guide to edible insects. Rochester VT, Park Street Press, USA.

Ramos-Elorduy J. 2005. Insects a hopeful food. In: M. Paoletti (Ed). Ecological implications of minilivestock. Oxford IBH Publishing, Enûeld, New Hampshire, USA.

Ramos-Elorduy J. 2008. Energy supplied by edible insects from Mexico and their nutritional and ecological importance. *Ecology of Food and Nutrition*. 47: 280–297.

Ramos-Elorduy J. 2009. Anthropo-entomophagy: Cultures, evolution and sustainability. *Entomological Research*. 39: 271–288.

Ramos-Elorduy J, de Morales LJ, Pino JM, Nieto Z. 1988. Contenido de tiamina riboflavin y niacin en algunos comestibles de Mexico. *Revista Tecnologia Alimentos*. 23:21. (In Spanish).

Ramos-Elorduy J, Gonzalez EA, Hernandez AR, Pino JM. 2002. Use of *Tenebrio molitor* (Coleoptera: Tenebrionidae) to recycle organic wastes and as feed for broiler. *Journal of Economic Entomology*. 95: 214-220.

Ramos-Elorduy J, Pino JM. 1990. Contenido calorico de algunos insectos comestibles de Mexico. Revista de la Sociedad Quimica de Mexico. 34: 56-68. (In Spanish).

Ramos- Elorduy J, Pino MJM, Escamillia PE, Alvadro PM, Otero JI, de Guevara OL. 1997. Nutritional value of edible insects from the state of Oaxaca, Mexico. *Journal of Food Composition and Analalysis*. 10: 142-157.

Ramos-Elorduy J, Pino MJM, Martínez CVH. 2008. Base de datos de insectos comestibles de México. *IBUNAM-UNIBIO*, Mexico. (In Spanish).

Ramos-Elorduy J, Pino JM, Rincon F, Marquez C, Escamilla E, Alvarado M. 1984. Protein content of some edible insects in Mexico. *Journal of Ethnobiology*. 4: 61-72.

Ramos-Elorduy J, Villegas J, Pino JM. 1989. The efficiency of the insect (*Musca domestica* L.) in recycling organic wastes as a source of protein. *Biodeterioration*. 7:805-810.

Raubenheimer D, Simpson SJ. 1990. The effects of simultaneous variation in protein, digestible carbohydrate and tannic acid on the feeding behaviour of larval *Locusta migratoria* (L.) and *Schistocerca gregaria* (Forskal). I. Short-term studies. *Physiological Entomology*. 5: 219-223.

Richards AL. 1939. Land labor and diet in Northern Rhodesia, an Economic Study of the Bemba Tribe. Oxford University Press, London.

Riddick CC. 1984. Health, aquariums, and the non-institutionalized elderly. *Marriage and Family Review*. 88: 163-172.

Rossiter MC. 1991a. Environmentally based maternal effects—a hidden force in insect population dynamics. *Oecologia*. 87: 288–294.

Rossiter MC. 1991b. Maternal effects generate variation in life history—consequences of egg weight plasticity in the gypsy moth. *Functional Ecology*. 5: 386–393.

Rossiter MC, Cox-Foster DL, Briggs MA. 1993. Initiation of maternal effects in Lymantria dispar—genetic and ecological components of egg provisioning. *Journal of Evolutionary Biology*. 6:577–590.

Ruddle K. 1973. The human use of insects: examples from the Yukpa. *Biotropica* 5:94-101.

Saalah S, Shapawi R, Othman NA, Bono A. 2010. Effect of formula variation in the properties of fish feed pellet. *Journal of Applied Sciences*. 10: 2537-2543.

Saha HK, Haldar P. 2009. Acridids as indicators of disturbance in dry deciduous forest of West Bengal in India. *Biodiversity and Conservation*. 18: 2343-2350.

Saha HK, Sarkar A, Haldar P. 2011. Effects of anthropogenic disturbances on the diversity and composition of the acridid fauna of sites in the dry deciduous forest of West Bengal, India. *Journal of Biodiversity and Ecological Sciences*. 1. e 1-11.

Sales J, Janssens GPJ. 2003. Nutrient requirements of ornamental fish. *Aquatic Living Resources*. 16: 533-540.

Sanchez NE. Onsager JA. 1988. Life history of *Melanoplus sanguinipes* (F) in two crested wheat grass pastures. *Canadian Entomologist*. 120: 39-44.

Santhanam R, Velayutham P, Jegatheesan G.1989. A manual of fresh water ecology (An aspect of fishery environment). Daya Publishing House, New Delhi.

Santos-Oliveira JF, de Passos CJ, de Bruno S, Magdalena SM. 1976. The nutritional value of four species of insects consumed in Angola. *Ecology of Food and Nutrition*. 5: 90-97.

Sathiadas R, Narayanakumar R. 1994. Price policy and fish marketing system in India. *Biology Education*. 11: 225-241.

Scriber JM, Slansky F. 1981. The nutritional ecology of immature insects. *Annual Review of Entomology*. 26: 183-211.

Seira AB. 1998. The use of golden snail (*Pomacea* sp.) as animal feed in the Philipines. *Voedings Magazine* 11: 40-43.

Sharma OP, Kishan R. 2006. Growth of common carp, *Cyprinus carpio* Var. *communis* (L.) fingerlings fed with various fermented meals. *Journal of the Inland Fisheries Society of India*. 38: 45-52.

Shim KF, Chua YL. 1986. Some studies on the protein requirement of the guppy *Poecilia reticulata* (Peters). *Journal of Aquariculture and Aquatic Sciences*. 4:79-84.

Singh OT, Chakravorty J. 2008. Diversity and occurrence of edible orthopterans in Arunachal Pradesh with a comparative note on edible orthopterans of Manipur and Nagaland. *Journal of Nature Conservation*. 19: 169–176.

Singh OT, Chakravorty J, Nabom S, Kato D. 2007a. Species diversity and occurrence of edible insects with special reference to coleopterans of Arunachal Pradesh. *Journal of Nature Conservation*. 19: 169–176.

Singh OT, Nabom S., Chakravorty, J. 2007b. Edible insects of nishi tribes of Arunachal Pradesh. *Hexapoda*. 14:56–60.

Singh PK, Gaur SR, Chari MS. 2006. Development of supplementary fish feed from low cost indigenous materials. *Journal of the Inland Fisheries Society of India*. 38: 26-30.

Sinha BRK. 1997. Environmental degradation of infancy stage at Santiniketan. In: Nag, V. K. Kumar, J. Singh. (Eds). *Geography and Environment:* Local issues.

Smith JL, Opekun AR, Larkai E, Graham DY. 1989. Sensitivity of the esophageal mucosa to pH in gastroesophageal reflux disease. *Gastroenterology*. 96: 683–689.

Sogbesan OA, Ugwumba AAA, Madu CT. 2006. Nutritive potentials and utilisation of garden snail (*Limicolaria aurora*) meat meal in the diet of *Clarias gariepinus* fingerlings. *African Journal of Biotechnology*. 5: 1999-2003.

Sogbesan OA, Ugwumba AAA. 2006. Bionomics evaluation of garden snail (*Limicolaria aurora*, Jay, 1937; Gastropoda: Limicolaria) meat meal in the diet of *Clarias gariepinus* fingerlings (Burchell, 1822). *Nigerian Journal of Fisheries*. 2: 358-371.

Sogbesan OA, Ugwumba AAA. 2008. Nutritional evaluation of termite (*Macrotermes subhyalinus*) meal as animal protein supplements in the diets of *Heterobranchus longifilis* (Valenciennes, 1840) Fingerlings. *Turkish Journal of Fisheries and Aquatic Sciences*. 8: 149-157.

Sonaiya EB. 1995. Feed resources for small holder poultry in Nigeria. *World Animal Review*. 82: 25-33.

Srivastava SK, Babu N, Pandey H. 2009. Traditional insect bioprospecting—as human food and medicine. *Indian Journal of Traditional Knowledge*. 8: 485–494.

Steinfeld H, Gerber P, Wassenaar T, Castel V, Rosales M, *et al.*, 2006. Livestock's long shadow; environmental issues and options. Food and Agriculture Organization of the United Nations, Rome.

Sugita H, Yamada S, Konagaya Y, Deguchi Y. 1999. Production of b-N-acetylglucosaminidase and chitinase by *Aeromonas* species isolated from river fish. *Fisheries Science*. 65: 155–158.

Taylor FA, Kuo FE. 2009. Children with attention deficits concentrate better after walk in the park. *Journal of Attention Disorders*. 12: 402-409.

Taylor RL. 1975. Butterflies in my stomach: Insects in human nutrition. Woodbridge Press Publishing Company, London.

Teotia JS, Miller BF. 1973. Fly pupae as a dietary ingredient for starting chicks. *Poultry Science*. 52: 1830-1835.

Teotia JS, Miller BF. 1974. Nutritive content of house fly pupae and manure recidue. *British Poultry Science*.15: 177-182.

Thorne PS. 2007. Environmental health impacts of concentrated animal feeding operations: Anticipating hazards - Searching for solutions. *Environmental Health Perspectives* 115: 296–297.

Timmerman CM, Chapman LJ. 2004. Hypoxia and interdemic variation in *Poecilia latipinna*. *Journal of Fish Biology*. 65: 635–650.

Turkey S, Szaboti RE. 1981. Dried bee meal as a feed stuff of grown turkeys. *Journal of Animal Science*. 61: 965-968.

Ugwumba AAA, Ugwumba AO, Okunola AO. 2001. Utilisation of Live maggot as supplementary feed on the growth of *Clarias gariepinus* (Burchell) fingerlings. *Nigerian Journal of Science*. 35:1-7.

Uvarov BP. 1966. Grasshoppers and locusts: A handbook of general acridology. Vol 1. Cambridge University Press, London.

Vohra P, Hilgurdia G, Williams CO. 1983. Establishing relationship of nutrient composition and quality of wheat and triticale grains using chick, quail and flour beetle bioassay. *Hilgardia*. 51: 1 -11.

Wade JW. 1992. The relationship between temperature, food intake and growth of brown trout, *Salmon trutta* (L) fed natural and artificial pelleted diet in earth ponds. *Journal of Aquatic Sciences*. 7: 59-71.

Wahla MA, Haq KA. 1976. Comparative consumption of *Sorghum* foliage by the adults of three grasshopper species. *Pakistan Journal of Agricultural Sciences*. 13: 155-158.

Waldbauer GP. 1968. The consumption and utilisation of food by insects. *Advances in Insect Physiology*. 5: 229-288.

Wang D, Zhai S, Zhang C, Zhang Q, Chen H. 2007. Nutritional value of the Chinese grasshopper *Acrida cinerea* (Thunberg) for broilers. *Animal Feed Science and Techonology*. 135: 66-74.

Yang GH. 1998. Utility of Chinese Resource Insects and Its Industrialization. China Agrilture Science Press, Beijing.

Yang HJ, Liu YJ, Tian LX, Liang GY, Lin HR. 2010. Effects of supplemental lysine and methionine on growth performance and body composition for grass carp (*Ctenopharyngodon idella*). *American Journal of Agriculture and Biological Science*. 5: 222-227.

Yen AL. 2009. Edible insects: Traditional knowledge or western phobia? *Entomological Research*. 39: 289–298.

www.ingramcontent.com/pod-product-compliance
Ingram Content Group UK Ltd.
Pitfield, Milton Keynes, MK11 3LW, UK
UKHW021011290726
14059UKWH00001BA/80

9 789386 071422